Observing Systems for Atmospheric Composition

Guido Visconti
Piero Di Carlo
William H. Brune
Mark Schoeberl
Andreas Wahner
Editors

Observing Systems for Atmospheric Composition

Satellite, Aircraft, Sensor Web and Ground-Based Observational Methods and Strategies

With 124 Illustrations

Springer

Editors

Guido Visconti
CETEMPS-Dipartimento di Fisica
Università degli Studi di L'Aquila
via vetoio
67010 Coppito (AQ), Italy
guido.visconti@aquila.infn.it

Piero Di Carlo
CETEMPS-Dipartimento di Fisica
Università degli Studi di L'Aquila
via vetoio
67010 Coppito (AQ), Italy
piero.dicarlo@aquila.infn.it

William H. Brune
Department of Meteorology
Penn State University
505 Walker Building
University Park, PA 16802, USA
brune@meteo.psu.edu

Mark Schoeberl
Earth Sciences Directorate
Mail Code 900
NASA Goddard Space Flight Center
Greenbelt, MD 20771, USA
mark.r.schoeberl@nasa.gov

Andreas Wahner
Forschungszentrum Jülich GmbH
ICG-II: Troposphäre
D-52425 Jülich
Germany
A. wahner@fz-juelich.de

ISBN-13: 978-1-4419-2147-5
e-ISBN-10: 0-387-35848-X
e-ISBN-13: 978-0-387-35848-2

9 8 7 6 5 4 3 2 1

springer.com

Preface

The goal of the third International Summer School on Atmospheric and Oceanic Science (ISSAOS 2004) was to bring together experts in observing systems and the atmospheric sciences to discuss the need for an observing system for atmospheric composition, its components, and the integration of components into a system. Much of the lecture material was conceptual, with the idea to provide attendees with a context to put their own component of the observing system.

The Local Committee, Guido Visconti and Piero Di Carlo, started to think about this school in the summer 2002 and asked William Brune to be a director. Prof. Brune accepted with enthusiasm and was able to get Mark Schoeberl and Andreas Wahner as co-directors. Because the director soon realized that they needed another year to put together all the speakers he had in mind, the school was held from 20–24 September 2004 in L'Aquila (Italy). The speakers were P. K. Bhartia, W. Brune, J. Burrows, J.-P. Cammas, K. Demerjian, H. Fischer, D. Jacob, P. Newman, K. Reichard, V. Rizi, M. Schoeberl, M. Schultz, U. Schumann, A. Thompson, C. Trepte, A. Wahner.

This edition of ISSAOS, for the first time, asked students to provide an evaluation of the school at its conclusion. The students generally liked the school, including the quality of the lectures, the opportunities to ask the lecturers questions, and accessibility of the lecturers for conversations. These results encouraged us to put together the lectures of the school in a book to give a larger audience the opportunity to learn about the observational and modeling techniques used to understand the atmospheric composition from satellites, aircraft, and ground-based platforms. For many lectures were two common themes: the role of each component in an observing system for atmospheric composition, and the advances necessary to improve the understanding of atmospheric composition.

For the school's organization we acknowledge the financial contribution of the Italian Ministry of the Environment, and of Center of Excellence CETEMPS.

We are also grateful to Simona Marinangeli and Manuela Marinelli for the help in the school's organization and the hard work of rearranging the lectures in an editorial format.

January 2006

G. Visconti
P. Di Carlo
W. H. Brune
M. Schoeberl
A. Wahner
(Editors)

Contents

List of Contributors

Dr. Pawan K. Bhartia
NASA Goddard Space Flight Center
Mail Code 916
Greenbelt, MD 20771
USA
e-mail: pawan.bhartia@nasa.gov

Prof. William H. Brune
The Pennsylvania State University
Department of Meteorology
505 Walker Building
University Park, PA 16802
USA
e-mail: brune@essc.psu.edu

Dr. Jean-Pierre Cammas
CNRS (UMR 5560)
Laboratoire d'Aérologie
OMP, 14 Avenue E. Belin
31400 - Toulouse
France
e-mail: camjp@aero.obs-mip.fr

Prof. Kenneth L. Demerjian
State University of New York at Albany
Atmospheric Sciences Research Center
251 Fuller Road
Albany, NY 12203
USA
e-mail: kld@asrc.cestm.albany.edu

Prof. Herbert Fischer
Institut für Meteorologie und Klimaforschung
Forschungszentrum Karlsruhe GmbH
Hermann-von-Helmholtz-Platz 1
D-76344 Eggenstein-Leopoldshafen
Germany
e-mail: herbert.fischer@imk.fzk.de

Prof. Daniel J. Jacob
Harvard University
Pierce Hall, 29 Oxford St.
Cambridge, MA 02138
USA
e-mail: djacob@fas.harvard.edu

Dr. Paul A Newman
NASA-Goddard Space Flight Center
Mail Code 916 Building 33, Room E320
Greenbelt, MD 20771
USA
e-mail: paul.a.newman@nasa.gov

Dr. Karl M. Reichard
The Pennsylvania State University
Applied Research Laboratory
P.O. Box 30
State College, PA 16804
USA
e-mail: kmr5@psu.edu

Dr. Vincenzo Rizi
CETEMPS-Dipartimento di Fisica
Università degli Studi di L'Aquila
Via vetoio
67010 Coppito-L'Aquila
Iytaly
e-mail: vincenzo.rizi@aquila.infn.it

Dr. Richard B. Rood
Earth and Space Data Computing Division
Mail Code 930
NASA-Goddard Space Flight Center
8800 Greenbelt Road
Greenbelt, MD 20771
USA
e-mail: Richard.B.Rood@nasa.gov

Dr. Mark Schoeberl
Earth Sciences Directorate
Mail Code 900
NASA Goddard Space Flight Center
Greenbelt, MD 20771
USA
e-mail: mark.r.schoeberl@nasa.gov

Dr. Martin Schultz
Biogeochemical System Department
Max-Planck-Institut für Meteorologie
Bundesstr. 55
D-20146 Hamburg
Germany
e-mail: martin.schultz@dkrz.de

Prof. Ulrich Schumann
Institut fuer Physik der Atmosphaere
DLR Oberpfaffenhofen
Postfach 1116
D-82230 Wessling
Germany
e-mail: Ulrich.Schumann@dlr.de

Prof. Anne M. Thompson
The Pennsylvania State University
Department of Meteorology
510 Walker Building
University Park, PA 16802
USA
e-mail: anne@met.psu.edu

Dr. Charles "Chip" R. Trepte
NASA Langley Research Center
Hampton, VA 23681-0001
USA
e-mail: c.r.trepte@larc.nasa.gov

Introduction I

Observing Systems for Atmospheric Composition

William H. Brune

The concept of observing systems is not new. Even at the dawn of weather satellites in the mid-1960s, planners were conceiving ways to combine the new cloud pictures from satellites with information from radiosondes, aircraft, and surface sites to improve weather forecasting. A 1965 drawing from the United States National Oceanic and Atmospheric Administration illustrates such an observing system.

Figure 1. Schematic of a fully integrated environmental monitoring system, circa 1965. (From the United States National Oceanic and Atmospheric Administration image archive.)

Even observing systems for atmospheric composition aren't new. Ozonesondes, aircraft, ground-based networks, field-intensive studies, and lidars have often been used together to advance the understanding of atmospheric composition. Satellites have been part of the observing system for stratospheric composition related to ozone and ozone changes for more than 30 years, but typically not part of the observing systems for atmospheric composition in the troposphere, the lowest layer of Earth's atmosphere.

So, why hold an International Summer School of Oceanic and Atmospheric Sciences on observing systems for atmospheric composition in 2004? We had three main reasons. First, the advances in computer technology, data distribution, collection, and assimilation, and modeling enable the use of information from observing systems in ways that were not possible before. Second, new satellite instruments can make global observations of key atmospheric constituents in the troposphere. Third, with the new computational and observing capabilities, understanding current global atmospheric composition and being able to predict future global atmospheric composition is becoming more of a reality.

Our planet is changing. A changing atmospheric composition is coupled with changes in other Earth components to determine the consequences of those changes. An observing system for atmospheric composition will thus help improve the understanding of and predictability for these other Earth components—climate variability and change, the water cycle, human contributions and responses, land use and land cover changes, ecosystems, and the carbon cycle—as well.

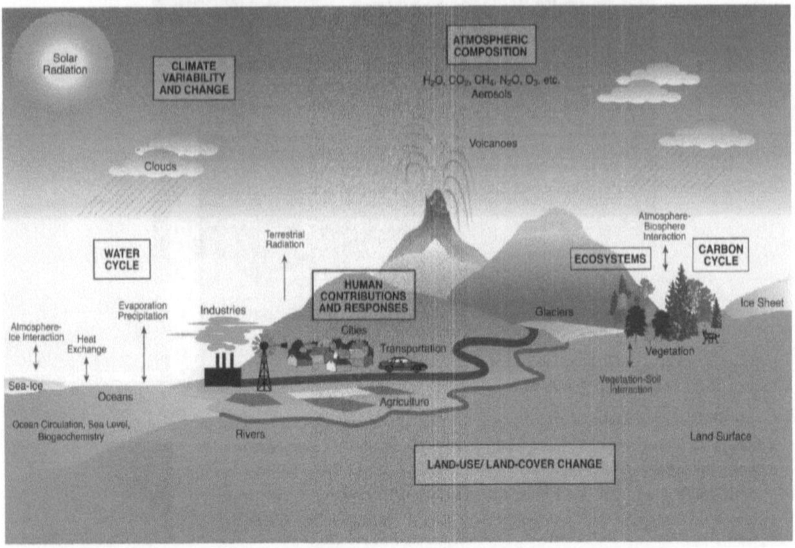

FIGURE 1. Schematic of the climate system (from *Our Changing Planet*, the U.S. Climate Change Science Program for Fiscal Years 2004 and 2005).

Introduction II

Needs for Sampling on Short Time and Spatial Scales

WILLIAM H. BRUNE

Introduction

While the major constituents of the Earth's atmosphere have not changed much since the advent of humans, the minor components have varied dramatically. Humans and their activities and natural processes both exert a powerful influence on atmospheric composition, sometimes with serious consequences. Understanding the variability and mitigating these consequences require an observing system for atmospheric composition. Components of an observing system for atmospheric composition have existed for decades, although recent advances in sensor technology and computational power make an integrated observing system more realizable. The observing system for atmospheric composition consists of instruments, models, and research.

The observing system is needed to develop an understanding of atmospheric composition and the processes that drive it and to provide the capability to confidently predict the interactions between atmospheric composition and changes and Earth and human activities. Two aspects of our existence are particularly affected by variations in atmospheric composition:

– air quality (human health, infrastructure and visibility degradation, ecosystem damage)

– climate (Earth's radiative balance and its cascading effects)

Thus, we are interested in the atmospheric composition that influences either atmospheric chemistry or Earth's radiation balance

One way to examine the observing needs for tropospheric composition is the list the issues affecting humans and the atmospheric processes that determine tropospheric composition (Figure 1). The issues range from the global scale to the regional and local scale: climate, atmospheric oxidation, global pollution, carbon balance, regional pollution and haze, and urban pollution and PM2.5 (particles $< 2.5\ \mu m$ in diameter). The processes are not all chemical: emissions

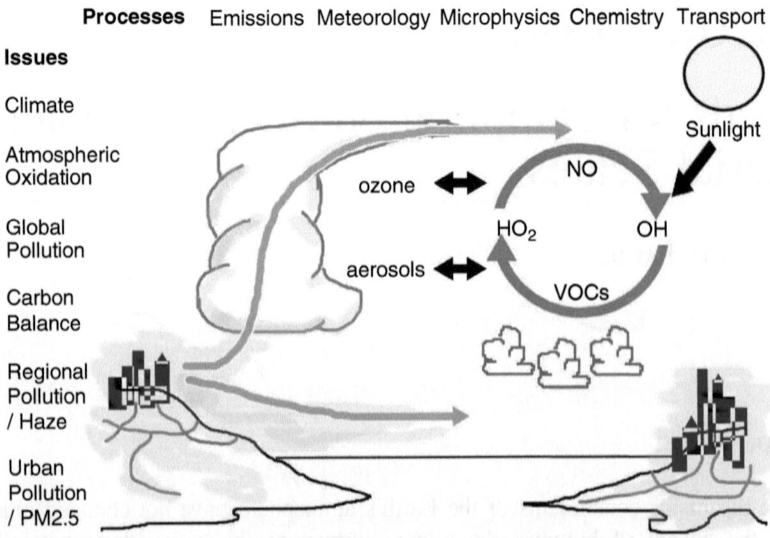

FIGURE 1. Schematic the processes and issues of atmospheric composition.

of atmospheric constituents from surface and airborne sources, meteorology, microphysics, chemistry, and transport.

We seek to answer two questions:

- How well can we describe these processes?
- Can we develop predictive capability for these issues?

The focus of this lecture will be on tropospheric composition and its behavior on short times scales and local to regional spatial scales. Of interest are water vapor (H_2O), carbon dioxide (CO_2), the pollutants ozone (O_3), and small particles and their precursors (particularly nitrogen oxides (NO_x=NO+NO_2), volatile organic compounds (VOCs), carbon monoxide (CO), sulfur dioxide (SO_2), and ammonia (NH_3), as well as acids and nitrates, toxic and carcinogenic gases. These atmospheric constituents can be either anthropogenic or natural. Even if we are only interested in pollution, we need to understand the interactions of pollutants with the environment.

Consider an atmospheric constituent that is a pollutant (e.g., an atmospheric component that has serious negative consequences for humans and their environment in trace amounts). Predicting the amount of this pollutant, which we will call A, requires observed variables, laboratory measurements of rate coefficients, and computer modeling.

Take the example of a simple box model. The time rate of change for an atmospheric constituent, A, can be determined from the equation:

$$\frac{d[A]}{dt} = \frac{q_A}{H} + P(A) - L(A)[A] - \frac{v_{dA}}{H}[A] + \frac{u}{\Delta x}([A]_o - [A]) \quad (1)$$

where
$[A]$ is the concentration of A (molecules cm^{-3})
q_A is the emission rate (molecules cm^{-2} s^{-1})
H is the height of the mixing box (cm)
$P(A)$ is the chemical production rate of A (molecules cm^{-3}s^{-1})
$L(A)$ is the 1st-order loss rate of A (s^{-1})
v_{dA} is the deposition velocity of A (cm s^{-1})
u is the horizontal wind velocity (cm s^{-1})
Δx is the width of the box in the wind direction (cm)
$[A]_o$ is the background value for $[A]$ (molecules cm^{-3})
At any given time, any one of these terms can be dominant, depending of the identity of A and the local meteorology. Often, several are comparable.

The amount of a pollutant that has adverse effects on humans and their environment depends on the atmospheric constituent and has changed over time as medical and public health studies improve. At present, the United States (EPA, 2004; WHO, 2000) and the World Health Organization have six pollutants whose adverse levels have been established (Table 1). Generally the ambient air-quality standards of the US, WHO, European Union, and other countries are similar, although there are differences.

Of these pollutants, CO, SO_2, and lead are primary emissions, NO_2 (nitrogen dioxide) is rapidly exchanged with NO (nitric oxide), which is a primary emission, particulate matter is both a primary emission and created by gas-to-particle conversion, while ozone is purely a secondary emission generated by atmospheric chemistry.

The emissions of greatest interest for local-to-regional scales and times of less than a month or so are those species with lifetimes less than that of CO (Figure 2). A goal of an observing system for atmospheric composition is to develop the capability to link species like CO, O_3, NO_x (=NO+NO_2) to the atmospheric radicals OH, HO_2, CH_3O_2, and NO_3, which are the driving forces behind the atmospheric chemistry that creates secondary pollution and modifies primary pollution.

TABLE 1.1. Asian Pollutant Emissions (Gg in year 2000) & 2σ Uncertainties

Country / Pollutant	China	%	Japan	%	India	%	Asia total	%
SO_2	20.400	13	800	9	5.540	26	34.300	16
NO_x	11.350	23	2.200	19	4.590	48	26.770	37
CO_2*	3.820	16	1.200	7	1.890	33	9.870	31
CO	115700	156	6.810	34	63.340	238	278560	185
CH_4	38.360	71	1.140	52	32.850	67	106820	65
NMVOC	17.430	59	1.920	35	10.840	149	52.150	130
BC	1.050	484	53	83	600	359	2.540	364
OC	3.390	495	74	181	2.840	544	10.420	450
NH_3	13.570	53	350	29	7.400	101	27.520	72

* CO_2 in Tg. From Streets et al., 2003

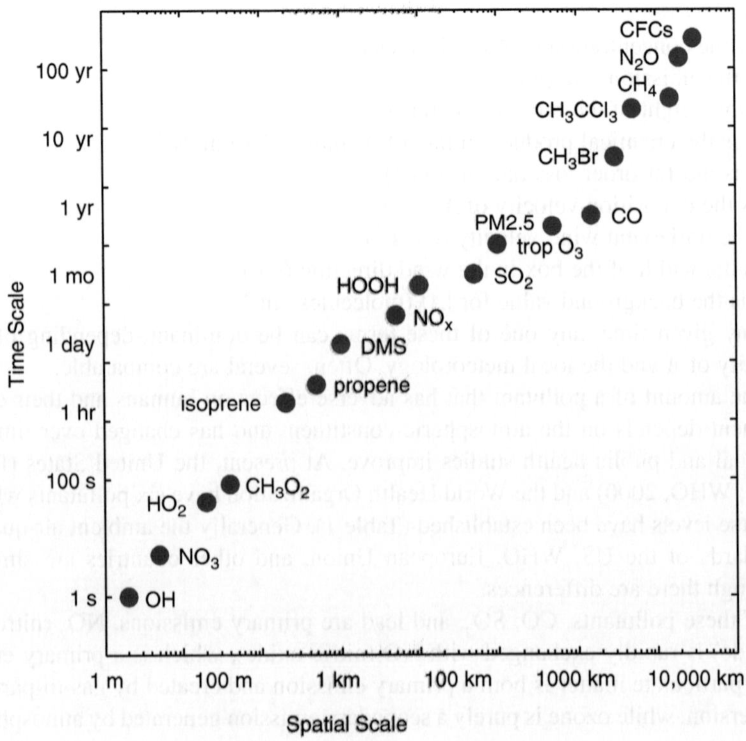

FIGURE 2. Spatial and temporal scales of atmospheric constituents.

Intensive studies show that the understanding of the processes that control atmospheric composition is reasonably good for some conditions. However, it does not yet enable the confident predictability needed for decision making regarding improved air quality, other socio-economic factors, and air-quality improvement deadlines. We examine the current understanding of each process in Figure 1.

Emissions

Governmental environmental agencies in the US and Europe have gone to great lengths to quantify a wide range of emissions. The result is that emission inventories are probably best in these two regions. In other regions the emission inventories are less well known. For instance, the inventory of gaseous and primary aerosol emissions in Asia in the year 2000 is fairly certain for atmospheric constituents like SO_2, NO_x, and CO_2, but is far less certain for CO,

TABLE 1.2. Emissions from Two Chemical Plants Near Houston, Texas

Plant	Emission	NO_x (kmole/hr)	C_2H_4 / NO (mole/mole)	C_3H_6 / NO (mole/mole)
Sweeney	inventory	14	0.01	0.01
	observed	15	3.6	2.0
	obs/inventory	1.07	360	200
Freeport	inventory	31	0.03	0.01
	observed	30	1.5	0.5
	obs/inventory	0.97	50	50

nonmethane volatile organic compounds (NMVOCs), and organic carbon (OC) and black carbon (BC) aerosols (Table 2).

Even in U.S. urban areas, inventories of some emissions are found to be significantly in error. An example is Houston, Texas, in September 2000, during the TexAQS2000 intensive field campaigns (TexAQS2000). The NOAA P-3 sampled plumes from two petrochemical plants on the coast south of Houston. In the plumes were highly elevated levels of ethane (C_2H_4) and propene (C_3H_6). By applying a transport model to the plume measurements, the emission rates were derived. The resulting emissions of NO_x agree with the emissions inventory (Table 3). However, the resulting ratios of ethene and propene to NO_x are more than 50–200 times larger than expected from the emissions inventory. The reasons for this difference are not clear, but it could be leaks or many small spillages that are less than the level that is required to be reported. What is clear is that the emissions inventories drastically underestimated the actual alkene emissions.

This error in the emissions inventory for ethene and propene is important for Houston's pollution chemistry. The hydroperoxyl radical, HO_2, is a precursor atmospheric constituent that leads to ozone production. It is mostly made during the day by processes involving sunlight, but is sometimes made at night, when ozone reacts with alkenes like propene and ethene. In Houston, a large nighttime propene spike leads to an HO_2 spike that is as large as HO_2 gets during the day (Figure 3). Thus, emission errors will propagate into errors in the modeling of radical chemistry and ozone production.

Conclusions regarding emissions:

- Some U.S. emissions are known quite well (e.g., NO_x from U.S. power plants); others are known quite poorly (e.g., ethene and propene).
- In the developing world, emission inventories are generally even less well known.
- There is also a factor of 2–3 uncertainty in regional natural emissions (such as for isoprene, a highly reactive VOC emitted by trees).
- Since emissions are the fuel for tropospheric (and stratospheric) photochemistry, it is important to know well the emissions, their distributions, and variations.

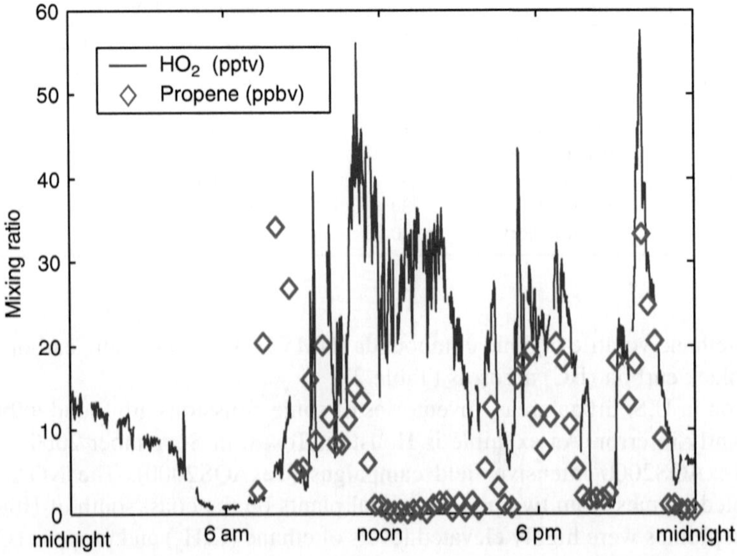

FIGURE 3. Propone emissions in the presence of ozone initiate fast atmospheric chemistry, as evidenced by the nighttime spike in propene and the reactive free radical HO_2 during the TexAQS2000 intensive study.

Tropospheric Chemistry

Where does ozone come from? In the stratosphere, ozone can result from the direct dissociation of molecular oxygen by ultraviolet light into oxygen atoms, which then join with molecular oxygen to form ozone. Ozone formation in the troposphere is more complex. It requires the presence of ultraviolet sunlight, although not as energetic as in the stratosphere, nitrogen oxides ($NO_x = NO + NO_2$), and volatile organic compounds (VOCs), which are the source of the free radicals, hydroxyl (OH), and hydroperoxyl (HO_2).

Ozone is a secondary pollutant—its only source is in the atmosphere. It is lost, though, both on surfaces and by photochemical reactions, as given in the following equation:

$$\frac{d[O_3]}{dt} = P(O_3) - L(O_3)[O_3] - \frac{v_{dO3}}{H}[O_3] + \frac{u}{\Delta x}([O_3]_o - [O_3])$$

where

$$P(O_3) = k_{HO2+NO}[NO][HO_2] + \sum_i k_{RO2i+NO}[NO][RO_{2i}]$$

$$L(O_3) = J(O_3)f_{H2O} + k_{HO2+O3}[O_3][HO_2] + k_{OH+O3}[O_3][OH]$$

Ozone losses include its photodestruction, followed by reaction with H_2O to form OH, direct reaction with OH and HO_2, and deposition on surfaces. Deposition and formation of OH are often the two largest ozone destruction processes.

The instantaneous ozone production, $P(O_3)$, comes from a reaction that cycles HO_2 to OH:

$HO_2 + NO \rightarrow OH + NO_2$
$NO_2 + \text{sunlight} \rightarrow NO + O$
$O + O_2 + M \rightarrow O_3 + M$

Ozone is also made by replacing HO_2 in the chemical equation above with RO_2, where R is CH_3, C_2H_5—a hydrocarbon with an odd number of hydrogen atoms. The instantaneous ozone production rate is determined to some extent by the cycling of HO_x between OH and HO_2 where $HO_x = OH + HO_2$ (Figure 4). Most HO_x is produced by photochemical processes as OH, although in environments where formaldehyde (HCHO) is high, a significant fraction of HO_x can come from the production of HO_2 by HCHO destruction by sunlight. Once created, OH is rapidly cycled to HO_2 by OH reactions with CO, CH_4, VOCs, and O_3 while HO_2 is cycled to OH by HO_2 reactions with NO and O_3. If the sources of OH and HO_2 were cut off, then OH would cycle to HO_2 in about a second. The reaction of OH with VOCs typically produces RO_2 before it produces HO_2. As a result, RO_2 abundances are roughly equal to HO_2 abundances. The amount of RO_2 and HO_2 produced is thus related to $P(HO_x)$ and the amount and type of VOCs that react with OH. When NO is greater than a few 10's of pptv, the reaction of HO_2 with NO dominates the reaction of HO_2 with O_3 and ozone is produced, not destroyed. Eventually, HO_x reacts to form more stable atmospheric constituents and the cycle is terminated. The reaction products depend on the HO_x

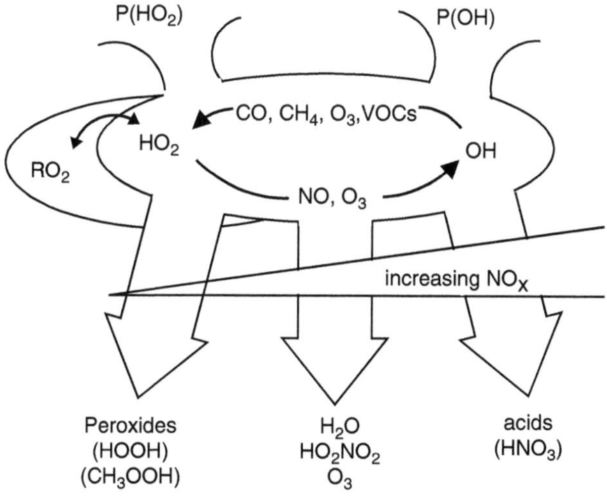

FIGURE 4. A schematic of the photochemistry of OH and HO_2 as a function of NO_x.

production rates (P(OH) and P(HO_2)), and on the NO_x abundances (Figure 4). Consider a case where P(HO_x) and the CO, CH_4, and VOC emissions are constant. At low NO_x, the HO_x is shifted to HO_2 (the HO_2/OH ratio is large) and the reactions of HO_2 with HO_2 and RO_2 produce the peroxides HOOH and ROOH. When NO_x is greater, more HO_x is shifted to OH (the HO_2/OH ratio is less) and the reaction of HO_2 with OH to form H_2O slightly dominates. When NO_x is very abundant, such as in a more polluted region or in the upper troposphere and stratosphere, then HO_x is shifted even more toward OH and the reaction of OH with NO_2 to form HNO_3 is most important. A similar diagram can be drawn for NO_x. In this case, the primary cycling between NO and NO_2 occurs by the reaction sequence:

$$NO_2 + \text{sunlight} \rightarrow NO + O$$
$$O + O_2 + M \rightarrow O_3 + M$$
$$NO + O_3 \rightarrow NO_2 + O_2$$
$$NO + HO_2 \rightarrow NO_2 + OH$$
$$NO + RO_2 \rightarrow NO_2 + RO$$

The first two equations show how NO_2 is converted into ozone. However, the reaction of NO with O_3 destroys O_3; a steady-state cycle forms during the day. No new O_3 is really created by this cycle. Instead, the O atom is merely exchanged between O_3 and NO_2. However, the reaction of NO with HO_2 and RO_2 does make new O_3.

The coupling of the HO_x and NO_x cycles produces O_3 (Figure 5). As the NO_x abundance is increased, OH first rises as more HO_x is shifted from HO_2 to OH and then falls, as OH is rapidly converted to HNO_3. HO_2 is relatively unaffected by increasing NO_x until NO_x reaches about 1 ppbv; then HO_2 decreases increasingly rapidly as HO_x is shifted from HO_2 to OH and as OH (and thus HO_x) is removed by NO_2. The resultant instantaneous ozone production rate thus increases until NO_x reaches a few ppbv and then decreases. Thus, instantaneous ozone production exhibits significant nonlinear behavior as a function of NO_x.

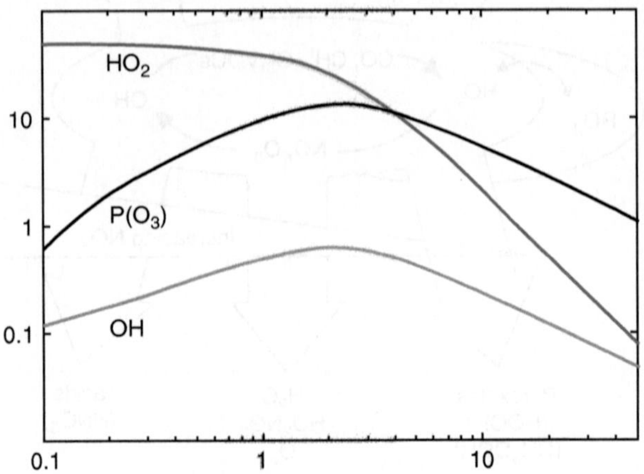

FIGURE 5. Variation of OH, HO_2, and P(O_3) as a function of NO_x.

An example of this nonlinear behavior can be seen in observations made during the TexAQS 2000 intensive field campaign in Houston, Texas (Figure 6). The behavior of NO, NO_2, and O_3 are shown for three days in August. Morning rush hour is marked by peaks in NO and NO_2, since NO is emitted by automobile and truck engines. Evening rush hour is also seen, although the peak values are smaller. However, it is clear that O_3 is building up over these three days, which means that O_3 production is occurring.

NO_x reaches several 10's of ppbv during rush hour. Thus, NO and HO_2 should anticorrelate. They do (Figure 7). However, since $P(O_3) \sim [NO]\,[HO_2]$, it can be large even when a small amount of HO_2 reacts with a large amount of NO, and vice versa. $P(O_3)$ remains between 40 ppbv/hr and 60 ppbv/hr from the hours of about 8:00 to 14:00. In this period, roughly 300 ppbv of ozone is created, which is proportional to the total ozone observed. Obtaining a more accurate ozone balance requires knowledge not only of the photochemistry, but also of the height of the mixed layer, the ozone loss on Earth's surface, and the ozone transport to and away from the region of interest.

The NO abundance is related to the NO_x abundance, and the RO_2 and HO_2 abundances are related to the VOC abundance. Because NO_x and VOCs are the primary emissions that contribute to ozone production, an ozone isopleth plot as a function of VOCs and NO_x abundance should provide guidance for the optimum regulatory actions to reduce ozone pollution. An early diagram (Figure 8) met the following conditions:

- It was chosen for a particular region, or box, which could be moving.
- The time duration of the run was chosen to capture the maximum ozone production.

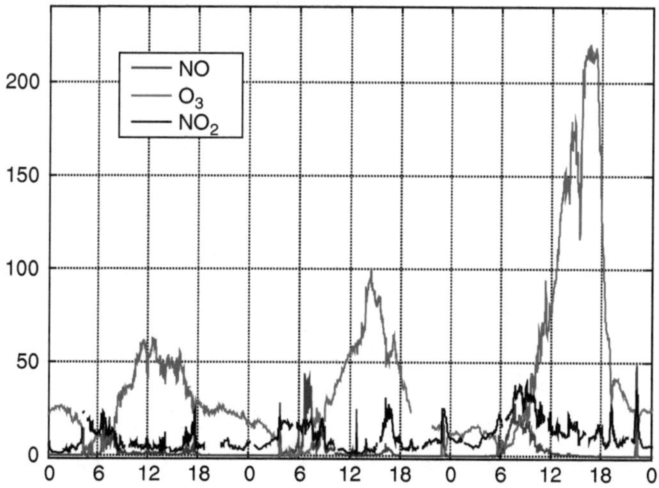

FIGURE 6. NO, NO_2, and O_3 during a pollution episode in Houston, Texas, during the TexAQS 2000 air-quality study.

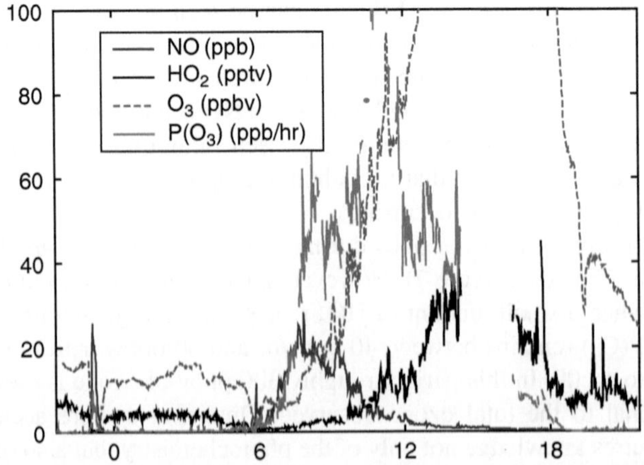

FIGURE 7. NO, HO_2, and $P(O_3)$ during a pollution episode in Houston, Texas, during the TexAQS 2000 intensive field campaign.

- The entire plot was generated by changing the VOC/NO_x mixture while leaving the meteorological conditions constant.

Regions above the ridgeline at $VOC/NO_x = 8$ are "VOC-limited"; more VOCs are need to produce more ozone. Regions on the plot below the ridgeline are "NO_x-limited"; more NO_x is needed to produce more ozone. Urban areas are typically

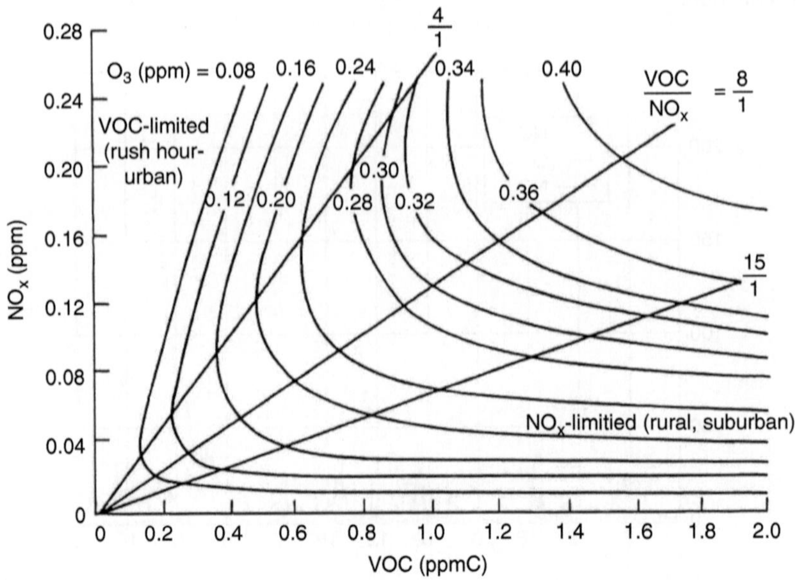

FIGURE 8. An example of an ozone isopleth diagram (adapted from NRC, 1990).

"VOC-limited," while rural and suburban areas and midday urban areas are often "NO_x-limited." This diagram indicates that reducing VOCs will result in reduced produced ozone, while reducing NO_x for areas above the ridgeline can result in an increase in ozone until the VOC/NO_x ratio drops below 8. This behavior results from the nonlinear coupling between HO_2 and NO.

Conclusions regarding atmospheric chemistry.

- The nonlinear behavior of HO_x–NO_x chemistry complicates the analysis of ozone production rates from observing systems that have low spatial and temporal resolution.
- Other products of tropospheric oxidation have effects on human health and climate, such as formaldehyde (HCHO), peroxyacetyl nitrate (PAN).
- The NO_3 radical plays a nighttime role similar to OH's daytime role.

Particle Microphysics and Chemistry

Particles have severe impacts on both human health and climate. Particles can be emitted directly from sources, such as diesel engines, or created in the atmosphere by gas-to-particle conversion. Particles less than 2.5 μm in diameter are particularly important for human health. They are small enough to navigate the respiratory system's passageways and are deposited deep within the lungs, where they can enter the bloodstream. PM2.5 has been linked to severe cardio-respiratory problems such as chronic asthma and bronchitis and even sudden heart attacks. At the same time, they are also important for Earth's radiation budget because they are about the size of the wavelength of the incoming solar radiation and are thus efficient at scattering or absorbing light.

In gas-to-particle conversion, volatile gases react with OH, O_3, or NO_3 to form gases that have low vapor pressures. These either condense on pre-existing aerosols or nucleate to form new ones. The main reactions involve SO_2, NO_2, VOCs, and NH_3:

$SO_2 + OH \rightarrow\rightarrow H_2SO_4 \rightarrow\rightarrow$ particles

$NO_2 + OH + N_2 \rightarrow N_2 + HNO_3 \rightarrow\rightarrow$ particles

$VOC_i + OH \rightarrow VOC_k$ (low vapor pressure) $\rightarrow\rightarrow$ particles

$NH_3 \rightarrow\rightarrow$ particles

The emission of primary particles and the gas-to-particle conversion of secondary particles lead to particle size distributions, which because of microphysics tend to have certain characteristics (Figure 9). Primary particles tend to be larger than 0.01 μm, while secondary particles are created at sizes in the 0.001–0.01 range when nucleation of low-vapor-pressure gases. Both types of particles grow by coagulation, condensation of both low-vapor-pressure gases and water vapor. These processes create particle size distributions: the nucleation mode (diameter < 0.01 μm), the Aiken mode (0.01 μm < diameter < 0.08 μm), the accumulation mode (0.08 μm < diameter < 2.5 μm), and the coarse mode ($d > 2.5$ μm). The nucleation and Aiken modes are rapidly lost by coagulation; the coarse mode is

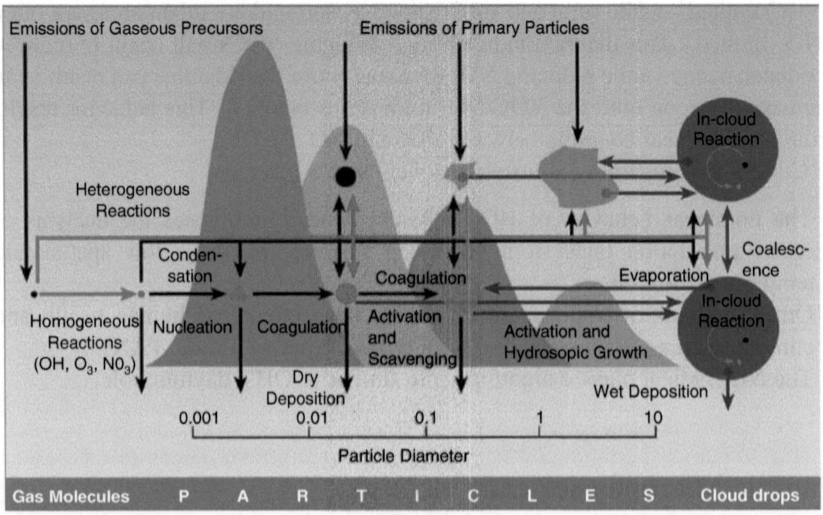

FIGURE 9. A schematic of atmospheric particles processes and size distributions (from US CCSP, 2004).

rapidly lost by deposition; the accumulation mode has a lifetime of a few weeks or so.

Equally important to the particle size distributions is the chemical composition. While no direct link has yet been made between the chemical composition of PM2.5 to human health, chemical composition has a major impact on climate. For example, some particles, such as sulfate (H_2SO_4), reflect incoming solar light, while other particles, such as soot from diesel exhaust, absorb incoming solar light. The impact on Earth's radiation balance is not trivial.

Small particles also have indirect effects on Earth's radiative balance. The nucleation of cloud drops and ice from pure water vapor is unrealistic in Earth's atmosphere. Instead, nucleation starts with particles. However, only a small subset of atmospheric particles are good cloud condensation nuclei and even a smaller subset are good ice nuclei. The chemical composition, along with the particle size, determines a particle's ability to be a cloud condensation nuclei (CCN) or an ice nuclei (IN).

In the troposphere, particles often contain a combination of sulfate, nitrate, organic or black carbon, and ammonia. The sulfate content in the U.S. East Coast is much greater than it is on the US West Coast, primarily because of the use of sulfur-containing coal in Midwest power plants. Organic content appears to be fairly ubiquitous, even in very remote regions of the Pacific Ocean.

Measurements were made with an aerosol mass spectrometer (AMS) in Scotland and Korea (Figure 10). In the Scotland sample, the size distribution is mostly in the Aiken mode and is dominated by the organic and nitrate components. In the Korean sample, the particle size distribution shows a growing accumulation mode and is dominated by the sulfate component. Peaks in the organic

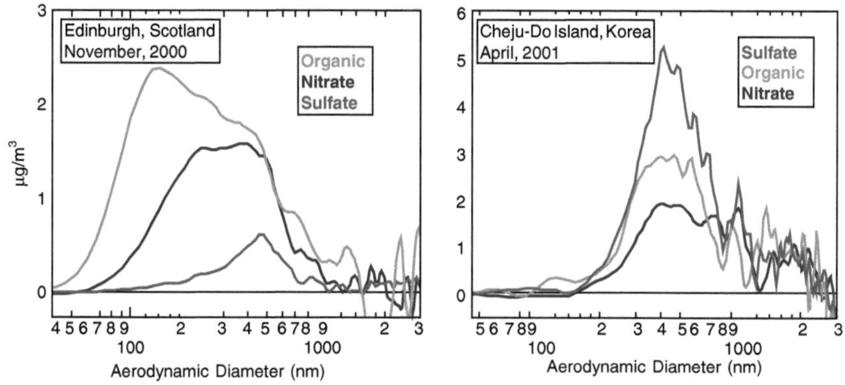

FIGURE 10. Particle size distributions and chemical composition in a city (Edinburgh, Scotland) and in a remote region downwind of urban and industrial regions (Cheju-Do Island, Korea). (from Hugh Coe, private communication, 2003).

component of the mass spectrum indicate that in Scotland the organic component consists primarily of hydrogen and carbon, while in Korea, the organic component has been oxygenated in the transit from the source regions. This oxidation process alters the particles' capability to act as CCN.

It was generally thought that there were always enough particles present in urban areas that low-vapor-pressure gases would always condense on pre-existing particles and not form new particles. However, recent measurements in Pittsburgh, PA, and other cities have shown that under certain meteorological conditions, such as recent frontal passages, new particles can nucleate (Figure 11). At 9 a.m., OH was created and reacted with SO_2 and VOCs to form new particles. This burst of new particles then rapidly coagulated, creating fewer, larger particles in a few hours. That new, small particles are formed may have implications for human health.

Conclusions for microphysics.

- Airborne particles participate in both air quality and Earth's radiative balance.
- Particle distribution usually consists of particles emitted directly from sources and particles created by gas-to-particle conversion.
- Particles' roles in human and health and climate are determined by both the particle size distribution and the particles' chemical composition
- Particles containing organics are widespread.
- $PM_{2.5}$ is a major research thrust—for reasons of both health effects and climate.

Meteorology

Meteorology affects the distribution of atmospheric constituents and atmospheric composition on spatial scales from turbulence to global transport and on temporal scales from seconds to decades. Without a knowledge of meteorology, it is

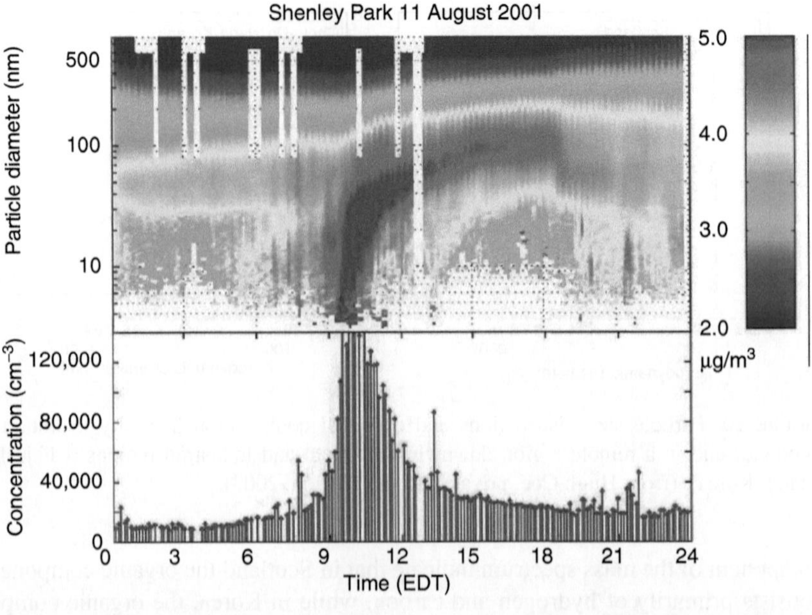

FIGURE 11. New particle formation & growth, 11/08/2001, Pittsburgh, PA. (From S. Pandis, private communication, 2003).

impossible to understand measurements of almost all atmospheric constituents, with the possible exception of the short-lived radicals CH_3O_2, NO_3, HO_2 in most conditions and OH in essentially all conditions (Figure 2). The OH lifetime is a second or less; its abundance and behavior are defined exclusively by reactions involving NO_x, VOCs, CO, and a few other constituents, and $P(HO_x)$. However, for OH, solar photolysis must be well known, and that depends on clouds. The interaction between meteorology and atmospheric composition continues to be an active research topic.

To illustrate the impact of meteorology on atmospheric composition, and its measurement, consider the planetary boundary layer, or PBL (Figure 12). During the day, surface heating generates convection, which raises and mixes surface gaseous and particulate emissions throughout the convective mixed layer, which is typically 1 to 4 km high and capped by a cloud layer. The circulation time constant is about an hour. While PBL constituents are vented into the free troposphere, especially by convective systems, the PBL can often be considered to be a box into which the surface emissions are mixed.

As sunset approaches, convection becomes less active and a lower, stable PBL forms. Atmospheric constituents that were higher in the convective mixed layer when convection ceased remain in the residual layer, disconnected to the surface. Any surface emissions occurring at night remain in the nocturnal boundary layer.

During rush hour near sunrise, high levels of vehicle exhaust, including NO_x, VOCs, and particulates, build up in the shallow boundary layer. As the sun heats

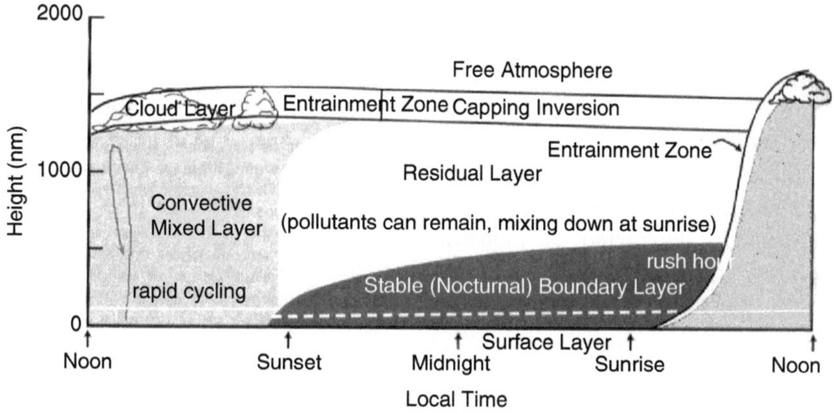

FIGURE 12. The typical behavior of the planetary boundary layer over the course of a day. (Adapted from Stull, Boundary Layer Meteorology, Kluwer, 1988).

the surface and convection begins, the air in the shallow boundary layer is mixed with air from the residual layer. This mixing increases until near midday, when the PBL height is now again greatest. Near the end of rush hour, surface emissions are mixed into a much larger volume, resulting in smaller pollutant mixing ratios since the mixing is into a greater volume.

The simultaneous pulse of pollutant emissions at morning rush hour and the increase in the PBL height after sunrise have important consequences. First, because the chemistry that produces ozone and particles is nonlinear with respect to the emitted pollutants' concentrations, the change in pollutant concentrations due to mixing into an increasing volume affects the production rates of ozone and particles. Second, the behavior of surface-based measurements of pollutant abundances can only be interpreted if the meteorology is known, particularly the variation of the PBL height.

The horizontal winds also strongly influence the distribution of pollution and atmospheric constituents, even on a neighborhood scale in urban areas. Chemical transport models transport atmospheric constituents from their emissions sources while transforming them with chemistry and microphysics. Urban airshed models use forecasts and analyses of meteorology on the few kilometer to synoptic scales (~1000 km).

Cloud-scale meteorology has two main effects. First, convective lifting of pollutants lifts local and regional pollutants to new heights, where they can be spread and where they can influence the photochemistry. Second, the process of going through clouds can scrub water soluble gases, changing the atmospheric composition and thus the chemistry.

Synoptic-scale meteorology creates the conditions necessary for pollution episodes. Stagnation episodes lead to smog (e.g., U.S. East Coast). Stable layers both create and inhibit regional surface pollution (e.g., Los Angeles, CA).

Synoptic-scale frontal systems lift pollutants and their precursors off the surface and into the free troposphere (e.g., Southeast Asia).

Global-scale winds move pollutants great distances from one region to another. There is growing recognition that in the Northern Hemisphere, pollution outflow from one continent is reaching the next continent downwind to the east. Thus, pollution from the United States gets to Europe; pollution from Europe gets to Asia; and pollution in Asia gets to the United States. Countries like Japan that are immediately downwind of industrialized and urbanized regions in China encounter more severe pollution transport. As developing countries in Asia industrialize, global pollution levels are likely to rise, with consequences for regions downwind.

Conclusions for meteorology.

- Meteorology on all scales influences the atmospheric composition.
- Because ozone and particle production can depend on constituent concentrations in a nonlinear way, the actual distribution of atmospheric composition can have a large impact on secondary pollutant production.
- The atmospheric composition of one region is influenced by the atmospheric composition of the upwind regions, even when those regions are thousands of kilometers away. Air quality has become a global issue.

Where to Go from Here?

Intensive measurements and monitoring of subsets of processes teach us a lot, but ...

- Which emissions, chemistry, microphysics, and meteorology are in common?
  – We need widespread, long-term process studies.
- Are observed differences due to differences in specific models and instruments?
  – We need frequent intercomparisons.
- Can calculations/observations of linear behavior be extrapolated, especially as the scientific basis for regulatory actions?
  – We need to understand atmospheric behavior over the entire range of variation; models must match both low and high values.
- Is "matching" ozone, or OH, or PM2.5 good enough?
  – We need to match all species (e.g., PAN, NO_x, etc.) and meteorology (e.g., PBL height, trajectories) that have importance influence on ozone and aerosols.
- Are convection and PBL mixing important?
  – We need to determine influence over the range of transport scales. We can't do chemistry without doing meteorology!
- What do we need to do to understand the problem?
  – We need to learn how to combine measurements from satellites, ground networks, intensive field campaigns, and models on several scales to build an integrated system for creating understanding and predictive capability.

References

U.S. Environmental Protection Agency (EPA), http://www.epa.gov/air/criteria.html, 2004.

National Research Council, *Rethinking the Ozone Problem in Urban and Regional Air Pollution*, National Academy Press, Washington, DC, 1991.

Our Changing Planet, The U.S. CCSP for fiscal years 2004 and 2005, pdf file at http://www.climatescience.gov, 2004.

Strategic Plan for the U.S. Climate Change Science Program, pdf file at http://www.climatescience.gov, 2003.

Streets, D.G., T.C. Bond, G.R. Carmichael, S.D. Fernandes, Q. Fu, D. He, Z. Klimont, S.M. Nelson, N.Y. Tsai, M.Q. Wang, J-H. Woo, and K.F. Yarber, An inventory of gaseous and primary aerosol emissions in Asia in the year 2000, *J. Geophys. Res.*, **108,** D21, 8809, doi:10.1029/2002JD003093, 2003.

WHO (World Health Organization), *Guidelines for Air Quality*. World Health Organization, Geneva, 2000.

Part I
Observations by Satellites

Chapter 1

Air-Quality Study from Geostationary/High-Altitude Orbits

PAWAN K. BHARTIA

Introduction

What determines air quality? There are two ingredients in the air that we breathe that in high concentrations cause poor air quality (AQ). They are particulates (aerosols) smaller than 2.5 μm, called PM2.5, and ozone. In addition, the US Environmental Protection Agency (EPA) regulates sulfur dioxide (SO_2), nitrogen dioxide (NO_2), and carbon monoxide (CO), which lead to enhanced concentrations of aerosols and O_3. SO_2 combines with water vapor to form sulfuric acid droplets that are the major constituents of urban smog. In the presence of sunlight, NO_2 photolyzes into NO and O; the latter combines with O_2 to form O_3. CO (as well as some volatile organic compounds) helps recycle NO back into NO_2; so the O_3 production continues until there is sunlight, or NO is converted by a competing reaction into HNO_3 and removed from the system.

There are serious health and environmental effects associated with high concentrations of aerosols and O_3. Small-size particulates can enter the lung causing breathing problems; they can also deposit cancerous materials, often found on the surface of these aerosols, into the lung. Enhanced concentrations of O_3 can affect the immune system and decrease agricultural productivity. However, it is not possible to control the amount of these pollutants simply by controlling the local emission of the source gases, for particulates and long-lived trace gases, such as CO and O_3, can be transported to large distances. For example, dust from the Sahara regularly blows north over Europe and is transported west to the Caribbean Islands and to the southeastern US by prevailing winds. Dust from China frequently crosses the Pacific to reach the northwestern US. Pollutants produced by biomass burning in Africa have been detected over some pristine Pacific islands. There are also natural events, such as forest fires, volcanic emission, etc, that can degrade AQ over large areas. Finally, while the AQ is obviously affected by weather and change in climate, there is also a reverse connection. Aerosols have several direct and indirect effects on cloud and radiation. They scatter sunlight and can increase cloud albedo. The carbonaceous aerosols (aerosols containing soot) reduce the amount of sunlight reaching the surface, thus suppressing the hydrological cycle. They also warm the atmospheric layer in which they are located and can suppress cloud formation.

Thus, the key scientific questions related to air quality one seeks to answer using satellite, ground-based, and aircraft instruments are

- What are the major sources of air pollution?
- How natural and anthropogenic processes interact in determining global air quality?
  - What are the relative influences of natural and manmade sources?
  - How are the pollutants transported across regions and continents?
  - How do radiation, weather, and climate affect air quality?
- How does air quality affect weather and climate?

Measurement of Air Quality by Satellite Instruments

The five pollutants mentioned above (O_3, SO_2, NO_2, CO, and PM2.5) are referred to as "criteria pollutants" by the US EPA. It turns out that all five can be measured by remote sensing instruments operating from satellites. In addition, satellite instruments can measure HCHO and HNO_3, both of which are related to the reactions that determine air quality, and finally, it is possible to estimate actinic flux in UV and visible, needed for the calculation of O_3 and NO_2 photolysis rates, from satellites.

The satellite instruments can be classified either based on the wavelength range in which they operate or based on their viewing geometry. The limb-viewing instruments, operating from visible to microwave wavelengths, are best for measuring particulates and trace gases in the upper troposphere (and above) with high sensitivity and good vertical resolution. However, in this lecture we will not consider such techniques; neither will we consider nadir-viewing thermal IR techniques for measuring trace gases in the middle and upper troposphere. This is because so far there has been very little experience in deriving AQ-related information from these techniques, except for CO. (This situation is likely to change in the coming months, as the data from the TES instrument, launched on July 15, 2004, on the EOS Aura satellite, are released.). In this lecture we will instead focus on the nadir-viewing techniques employed by the TOMS, GOME, and SCIAMACHY satellite instruments to measure the pollutants in the lower troposphere using reflected/scattered sunlight in UV, visible, and near-IR wavelengths (0.3–2.5 μm). The TOMS measurement series started in November 1978, and the GOME series in July 1995. SCIAMACHY employs techniques similar to GOME but provides higher spatial resolution and measures in the near-IR to retrieve CO column amounts. The following provides overviews of the techniques used by these instruments to measure the 5 criteria pollutants from space.

Aerosol Measurement

To study AQ one needs the surface concentration and size distribution of aerosols. In addition one needs their vertical distribution and absorptive properties to do photolysis calculations. Although the satellites cannot directly measure these quantities, they can provide important information to test, improve, and constrain the chemical and transport models that can provide the needed information. In the future, it may be possible to use data assimilation techniques, in which one statistically combines information in models and measurements, to produce a consistent three-dimensional aerosol field.

For decades aerosol optical depth has been derived from the measurement of top-of-the-atmosphere reflectance (TOAR) at a single wavelength in the visible. Figure 1 provides a simplified explanation of how the technique works. (The symbol ρ, commonly used for TOAR, will be defined in Figure 5 of page 53 on the derivation of total O_3 from the buv technique. It is a convenient way to convert the radiance measured by a satellite instrument into a dimensionless quantity that behaves like reflectivity.) From Figure 1, in absence of aerosols ρ becomes equal to the surface reflectance R; aerosols scattering (2nd term) increases ρ, while the 3rd and the 4th terms, which represent loss of sunlight to space by aerosol backscattering, and conversion into heat by aerosol absorption, respectively, can make ρ smaller than R for large R. If R is small, one can derive aerosol scattering optical depth (τ_s) from the 2nd term by assuming the aerosol scattering phase function $P(\Theta)$. Figure 1 shows several techniques for assessing $P(\Theta)$. The MODIS instrument flying on EOS Terra and Aqua satellites makes multispectral measurements of ρ to determine if the aerosols consist of small or large particles, which helps in the determination of $P(\Theta)$. The MODIS also assumes a value for the aerosol single scattering albedo (ratio of τ_s and τ_{ext}) to convert τ_s into aerosol

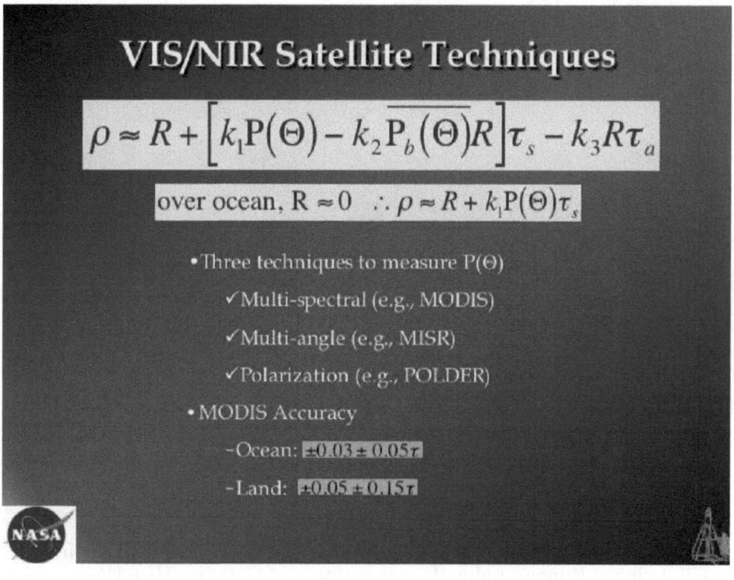

FIGURE 1. A simplified formula for the top-of-the-atmosphere reflectance measured by satellite instruments at visible/near-IR wavelengths.

extinction optical depth, τ_{ext}. Figure 1 shows an estimate of the accuracy in deriving τ_{ext} from MODIS [Remer et al., 2004]. One notes that the MODIS retrieval accuracy is significantly worse over land; over bright surfaces, e.g., deserts, MODIS is unable to retrieve any aerosol information. (This is because the surface reflectivity of land is more variable, and the last two terms in the equation in Figure 1 reduce the aerosol signal over bright surfaces.). Recently, a new algorithm, called the "deep-blue" algorithm, has been developed [Hsu et al., 2004] to improve the estimation of τ_{ext} over land. This algorithm uses the 412- and 490-nm MODIS channels where the land is darker than at the longer wavelengths. Figure 2 shows the τ_{ext} and Angstrom coefficient (a rough measure of aerosol size) derived from this algorithm using data from the SeaWIFS instrument, which has similar channels.

Finally, we describe a technique to estimate aerosol absorption optical depth τ_a using measurements of ρ at two ultraviolet wavelengths. The physical basis of this method is shown in Figure 3. Note the formula in Figure 3 is very similar to that in Figure 1, except the surface reflectivity is now replaced with ρ_0, the TOAR in absence of aerosols. In ultraviolet, ρ_0 is usually much larger than the surface reflectivity and has strong wavelength (λ) dependence caused by Rayleigh scattering from the molecular atmosphere. This leads to a curious

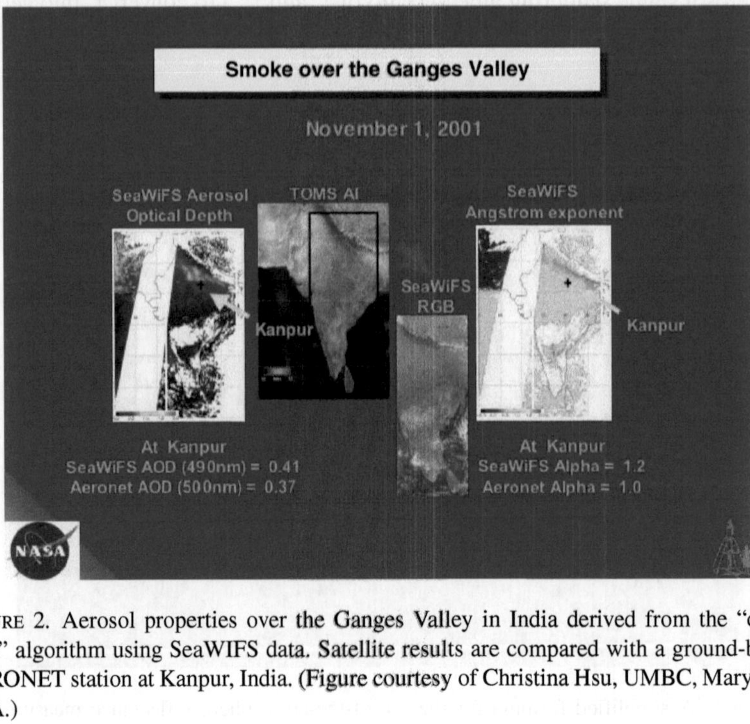

FIGURE 2. Aerosol properties over the Ganges Valley in India derived from the "deep-blue" algorithm using SeaWIFS data. Satellite results are compared with a ground-based AERONET station at Kanpur, India. (Figure courtesy of Christina Hsu, UMBC, Maryland, USA.)

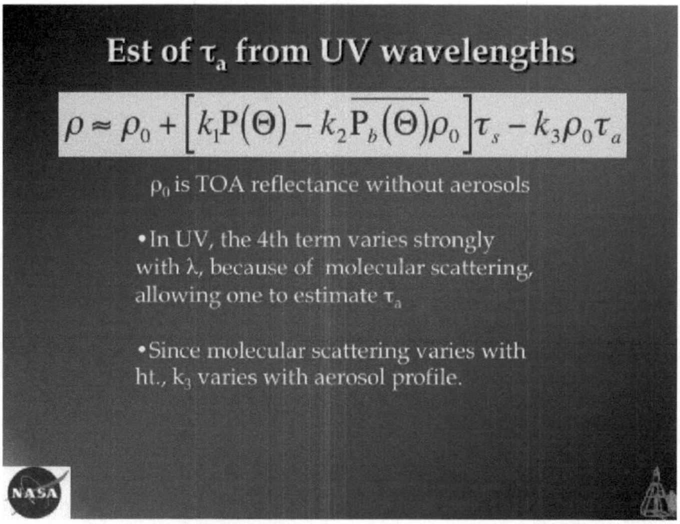

FIGURE 3. A simplified formula for the top-of-the-atmosphere reflectance measured by satellite instruments at UV wavelengths.

and a highly nonintuitive result: when τ_a is nonzero and significant (i.e., when significant amount of UV-absorbing aerosols, such as smoke and desert dust containing iron oxide are present), λ-dependence of ρ_0 causes the 4th term to have λ-dependence, which is over and above the λ-dependence of τ_a itself and modifies the λ-dependence of the 2nd term. For example, for smoke, which consists of small particles, τ_s increases with decrease in λ, but since τ_a is nearly independent of λ, the 4th term also does the same with λ, thus $\rho-\rho_0$ can either decrease or increase with λ depending on the relative magnitudes of the various terms. Generally speaking, k_3 is nearly zero for aerosols near the surface but increases as aerosols get elevated, so spectral dependence of $\rho-\rho_0$ can change sign depending the height of the smoke. By contrast, τ_2 for desert dust is independent of λ, but since τ_2 increases with decrease in λ, $\rho-\rho_0$ decreases with λ, irrespective of the height of the dust layer. These features have led to a unique method of detecting UV-absorbing aerosols from spectral contrast of measurements of $\rho-\tilde{\rho}_0$ at two UV wavelengths. This so-called aerosol index method has been applied to the TOMS data to study the transport of dust and smoke over the globe [Hsu et al., 1996; Herman et al., 1997]. A notable feature of the technique is that, since the aerosol index over cloud is zero, clouds do not show up in the aerosol index maps at all (Figure 4) unless the aerosols are above or embedded in the cloud. Indeed, the aerosol index method is the only (passive) remote sending technique that can detect (UV-absorbing) aerosols over clouds (Figure 5) as well as over very bright snow/ice covered surfaces.

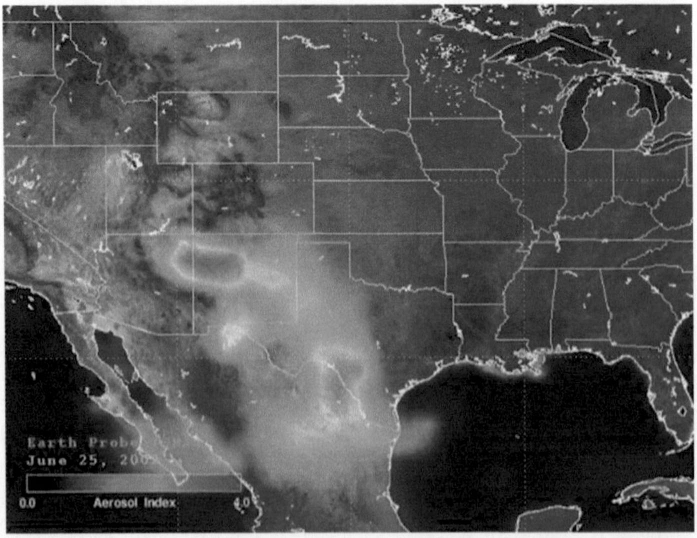

FIGURE 4. This false color map of TOMS aerosol index (AI) shows smoke from fires in the western US. AI is produced from the spectral contrast of $\rho-\rho_0$ between two TOMS wavelengths. The color of the land in the background is a visible RGB image from another sensor. Note the absence of clouds in the picture; the TOMS AI algorithm is not sensitive to clouds.

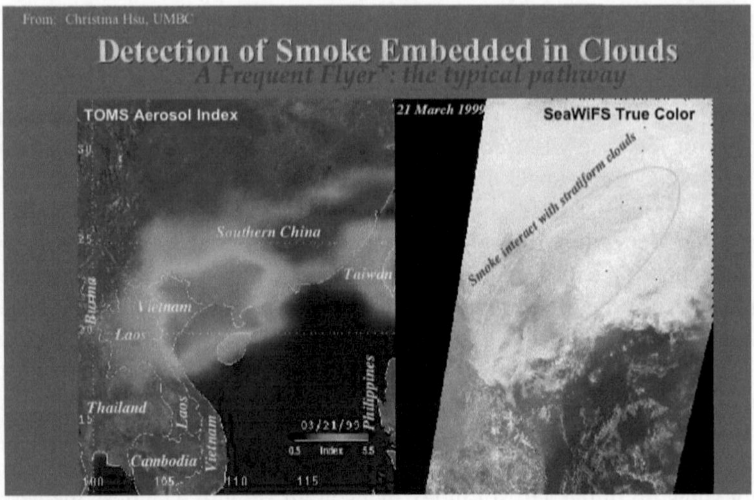

FIGURE 5. Right side of the figure shows a true color map of the SE China showing a thick cloud cover. The left side of the figure shows a TOMS AI map over the same region showing the presence of smoke, which is embedded in the cloud or above it. Upon close examination the right side of the figure shows that the clouds under the yellow oval are indeed darker, confirming the TOMS finding.

Using aerosol models and the Mie scattering theory it is possible to make quantitative estimates of τ_a and τ_s from measurements of ρ at two ultraviolet wavelengths [Torres et al., 1998]. With future instruments, the visible, the deep blue, and the UV methods can be combined to make estimates of both the scattering and absorption optical depths of aerosols over both land and ocean. Currently, the largest uncertainty in estimating τ_a comes from the fact that k_3 in Figure 3 is dependent on aerosol vertical distribution. To improve this estimate one needs space-based aerosol lidars.

Estimation of Tropospheric O_3

Estimation of tropospheric O_3 remains a difficult challenge for satellite instruments, for the radiation that is sensitive to the tropospheric O_3 has to pass through the stratosphere before reaching the satellite. The absorption of this radiation due to stratospheric O_3 is typically 20 or more times greater than due to tropospheric O_3. Moreover, the radiation generated in the troposphere is multiply scattered, which smears out the profile information. Finally, molecular and cloud scattering can significantly reduce the sensitivity to O_3 in the lower troposphere. So far there is no evidence that any satellite technique can sense the O_3 near the surface.

The best that satellite instruments have been able to achieve so far is to estimate the tropospheric O_3 column from total column O_3 by subtracting the stratospheric O_3 overburden. There are several ways of doing so, including the use of limb-viewing instruments, such as MLS and SAGE, that can measure down to the tropopause to estimate the stratospheric O_3 overburden [Fishman et al., 1990; Ziemke et al., 1998].

Here we describe a novel method, called the "convective cloud difference" (CCD) method in which total O_3 measurements made above high altitude clouds are used to estimate the stratospheric overburden [Ziemke et al., 1998]. The principle of the method is illustrated in Figure 6. Based on Kley et al. [1996] the deep convective clouds that often form in the Pacific Ocean pump very clean marine boundary layer air into the upper troposphere making the mixing ratio of O_3 in tropical transition layer (TTL), the layer between the top of these clouds and the tropopause, very small. Indeed Kley et al. sometimes detected near-zero ozone mixing ratio in the TTL. This means that the Pacific TTL often contains less than 1 DU of O_3, thus providing a very nice way of measuring the stratospheric column O_3 by measuring the column O_3 above very high deep-convective clouds.

Figure 7 shows comparison of the stratospheric O_3 column derived using the CCD method with those derived by integrating the O_3 profiles measured by the SAGE instrument.

SAGE is a solar occultation instrument that provides O_3 profiles with 1 km vertical resolution from tropopause to 60 km. Note that due to its limited sampling, SAGE results shown are zonal averages obtained from about 30 measurements in a month. The CCD measurements are from the Pacific only. (The implicit assumption here is that the stratospheric ozone column is longitudinally invariant.)

Nonetheless, there is a striking agreement between these two independent measurements. Given these measurements one can then estimate tropospheric O_3

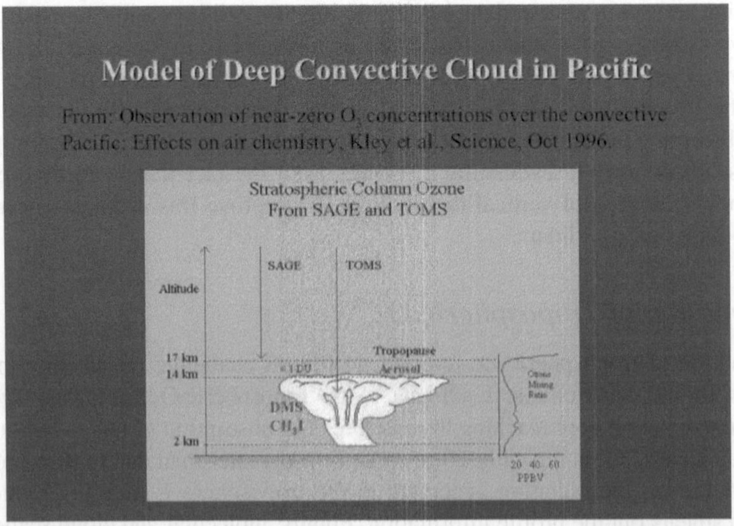

FIGURE 6. The tropical deep convective cloud model proposed by Kley et al. [1996] to explain their observation of near-zero ozone mixing ratio in the tropical transition layer (TTL). These clouds allow a simple yet powerful method of estimating stratospheric O_3 overburden that can be subtracted from total column O_3 (measured over cloudfree scenes) to study tropospheric column O_3.

column from total column O_3 measured over clear scenes. Figure 8 compares the times series derived this way with the time series from the balloon instruments measured at two tropical locations. Ziemke et al. [2001] have extended this method to a more general "cloud slicing" method in which clouds at multiple heights in the atmosphere can be used to estimate mean mixing ratio of O_3 in the upper and lower troposphere.

Nitrogen Dioxide (NO_2)

NO_2 is an inevitable byproduct of combustion, produced by automobiles and biomass burning; it is also produced by lightning. Though, by itself, NO_2 is not considered harmful, it is the primary source of atomic oxygen in the troposphere that leads to ozone production. The mixing ratio of NO_2 in the troposphere varies from less than 0.1 parts per billion to more than 10 per billion in highly polluted urban areas. However, since most of the NO_2 is vertically localized, the total column NO_2, typically measured by satellites, rarely exceeds few times 10^{16} molecules/cm^2, producing less than 1% reduction in TOAR (with respect to the background value) at ~400 nm, where NO_2 has maximum absorption. To make matters worse, from the point of view of satellite instruments, highest concentration NO_2 usually occurs in areas of less than 100 km^2 and remains localized in the planetary boundary layer (PBL) and is thus easily obscured by clouds. Yet,

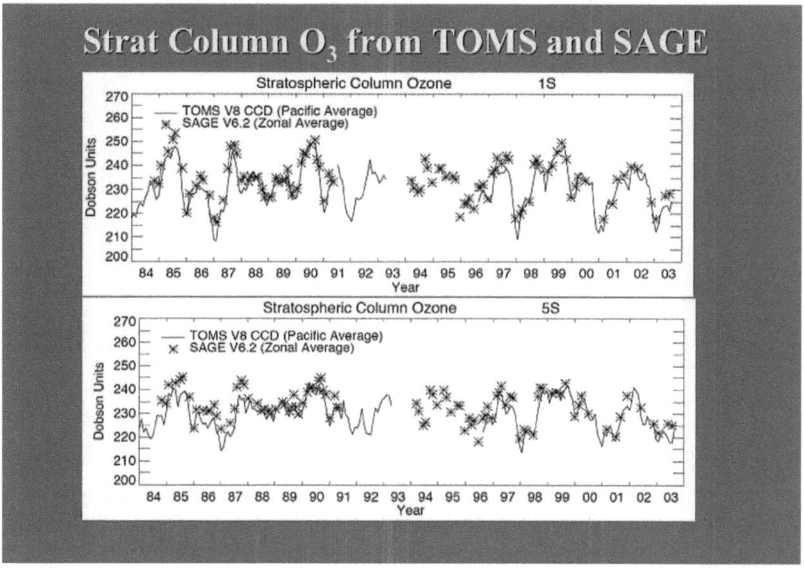

FIGURE 7. Twenty-year time series of stratospheric column O_3 derived using Pacific deep-convective clouds (CCD method) is compared with SAGE measured column O_3. These two independent methods agree remarkably well; both capture the very significant difference in behavior of the O_3 time series between 1 °S (upper panel) and 5 °S.

remarkably, it has been possible to detect NO_2 from satellite instruments [Richter & Burrows, 2002; Wenig et al., 2004]. Figure 9 shows a remarkable global map produced from data collected by the SCIAMACHY instrument on the ENVISAT satellite, showing the NO_2 pollution generated by all the major urban centers of the world. Despite these noteworthy successes in producing such maps, the accuracy and precision of the derived product remain marginal for careful quantitative studies of AQ. From the measurements made so far it is particularly difficult to track the NO_2 plumes away from the source regions; such transports are responsible for converting a local air pollution problem into a regional and even a global problem.

Carbon Monoxide (CO)

In principle, CO can be measured using the 2.3-micron absorption band where there is still a sufficient amount of sunlight to measure the reflected radiation. However, this band is extremely weak and so far the experience in using this band has been limited. (There are preliminary indications that the SCIAMACHY instrument is able to see CO using this band.). So far, satellite measurements of CO have come primarily from the 4.6-micron thermal emission band of CO,

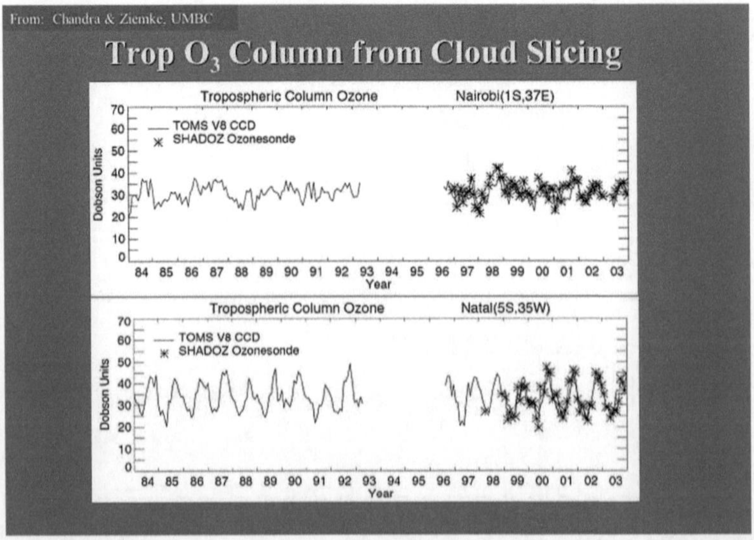

FIGURE 8. Twenty-year time series of tropospheric column O_3 derived using the two tropical locations. The CCD methods capture the very significant difference in behavior of the trop O_3 time series between Nairobi (upper panel) and Natal. These results give confidence that clouds can be used as a powerful tool to "slice" the atmosphere to study tropospheric O_3.

which provides information about CO in the middle troposphere. Recent results have come from the MOPITT instrument on the EOS Terra and the AIRS instrument on the EOS Aqua satellite.

Sulfur Dioxide (SO_2)

There are both natural and anthropogenic sources of SO_2. Large amounts of SO_2 are routinely emitted by fuming volcanoes, but they are short-lived and cause localized pollution. Volcanic eruptions can inject SO_2 into the upper troposphere and stratosphere where they quickly convert to sulfuric acid aerosols that may last a long time, even years, causing perturbation to climate and radiation. These high-altitude injections of SO_2 are fairly easy to monitor from satellites. Even the Nimbus-7 TOMS instrument, which was not designed to measure SO_2, but was later found to have weak sensitivity to it, has seen SO_2 injected into the atmosphere by about one hundred volcanic eruptions during its 13.5-year lifetime [Carn et al., 2003]. The more recent instruments—GOME, SCIAMACHY and OMI—have more than an order of magnitude greater sensitivity to SO_2 than TOMS. In particular, the recently launched OMI instrument, which has a smaller pixel size than its predecessor instruments, should be able to monitor the fuming volcanoes.

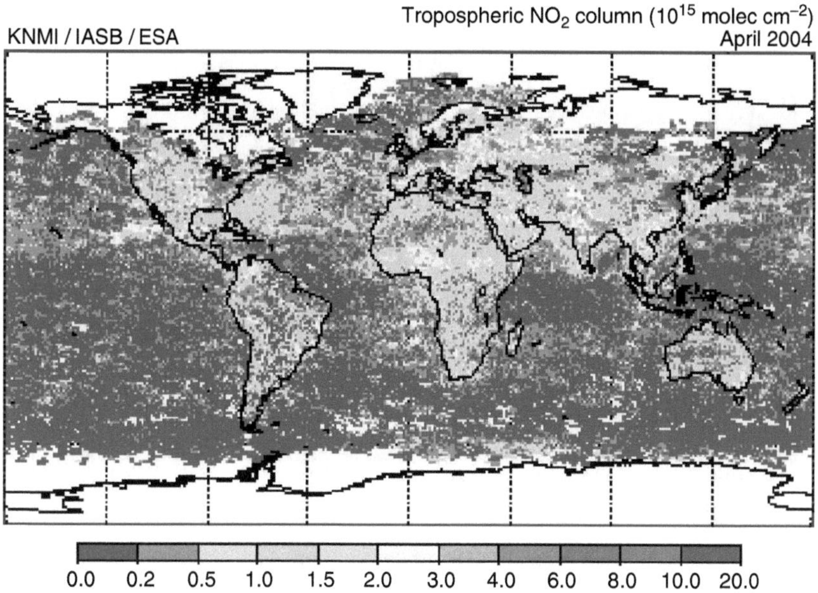

FIGURE 9. April 2004 map of tropospheric NO_2 produced from data collected by the SCIAMACHY instrument. (Map courtesy of KNMI, NL.)

The second major source of SO_2 is the burning of sulfur-rich coal. Although many industrialized nations have reduced the emission of SO_2 considerably in the last few decades by using sulfur scrubbers in their power plants, many countries in the developing world do not do so. In addition, in many countries, burning of sulfur-rich coal in homes and small factories are major contributors of SO_2. However, given the very short lifetime of SO_2 near the surface, with rare exceptions [Carn et al., 2004], anthropogenically generated SO_2 doesn't usually build up in large enough amounts to be seen easily from satellite instruments. Yet, as Figure 10 illustrates, it has been possible to measure SO_2 near the surface from satellites in some parts of the world where the SO_2 pollution is most severe. However, to increase the scientific value of such measurements it will be necessary to increase the sensitivity of these measurements by an order of magnitude.

Measurement from High Vantage Point Satellite Orbits

With the exception of aerosols, satellite measurements of atmospheric pollutants have been largely confined to the Low-Earth Orbit (LEO), roughly defined as orbits with altitude of less than 1000 km from the surface of the Earth. These orbits have several inherent problems for AQ studies. To measure over a ground-swath of, say, 1500 km, from a typical 750-km altitude orbit, requires a camera of field-of-view (FOV) of about 90° (assuming one is using a two-dimensional CCD camera, similar to the popular digital cameras of today). In 35-mm camera

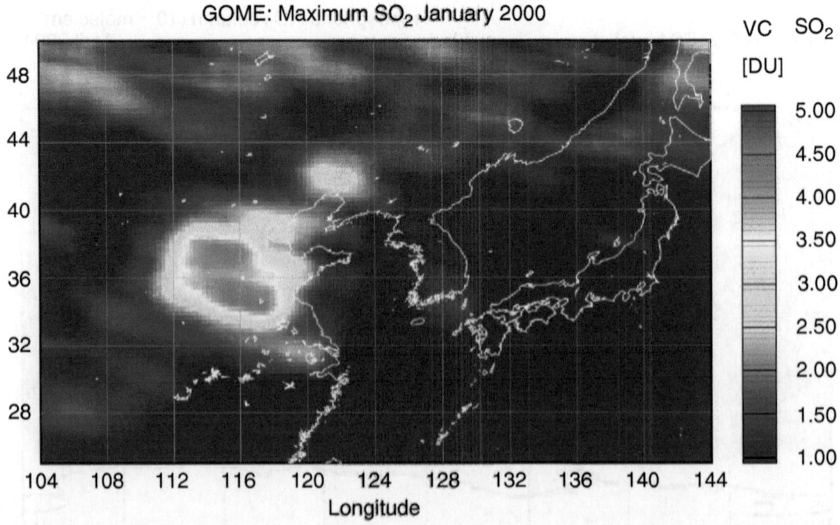

FIGURE 10. January 2000 map of SO_2 produced from data collected by the GOME instrument. (Map courtesy of Prof. John Burrows, U. of Bremen, Germany.)

terms, this amounts to using a 17.5-mm wide-angle lens. To get decent optical performance from such wide-angle optics, one is forced to use very small apertures (called the f/number), which limits the amounts of photons that can be collected. A typical aperture size for such a camera might be ~2 mm. Another serious problem with LEO instruments is that the satellite moves too fast (~7 km/sec), which severely limits the integration time ("shutter speed"); similar to taking the picture of a nearby object from a fast moving vehicle without blurring the picture. Finally, an LEO instrument (measuring reflected sunlight) with 1500-km swath will see any point on the Earth's surface once every two days, and perhaps every 4 days if cloud free pixels are required. Although the coverage can be improved by increasing the swath, it would be of little value for AQ studies for the pixels at the edges of the swath become much too large and are frequently cloud contaminated. Since AQ changes on a hourly scale, the temporal coverage provided by LEO instruments are deemed inadequate to study AQ.

Moving farther away from the Earth brings in several advantages. To start with, let's first dispose off the myth, held by many, that moving away from the Earth would reduce the signal/noise. Though the number of photons go down by the square of the distance, as users of SLR cameras know well, the number of photons received by a camera is a function of the f/number and not of its distance from the object; the distance effect is canceled exactly by the fact that longer focal length lenses of the same f/number have larger apertures. As we discussed above, the LEO instrument actually require a slower optics (larger f/number) to minimize optical distortion, which is not a problem for measurements made from larger distances. For example, if one were to move 40 times away from Earth, from 750 km to 30,000 km, one would require a modest 80-mm aperture instead of the 2-mm aperture (forced by the wide angle optics) to get exactly the same swath and S/N. And, since the FOV of the instrument would also decrease by a factor of 40, one could easily design a faster optics to improve S/N. Moreover, by the Kepler's law, satellites in more distant orbits move slower, so one can use a slower shutter speed, increasing the integration time to achieve even better S/N. Finally, and perhaps most importantly, a point on the Earth remains visible longer the farther one goes away, so one can pick and choose areas one wants to study and then study the area for long periods of time. One could, for example, wait for clouds to move away from an area to see what is underneath or for clouds to come over an area to isolate what is above them. Indeed the cloud slicing becomes a very compelling tool for sounding the troposphere from a high vantage point orbit.

The table in Figure 11 shows that there are many tradeoffs in choosing a high vantage point orbit. The primary one is the observation time versus spatial coverage.

Orbit Scenarios

	LEO	MEO	GEO	L1
Altitude (km)	564	20,187	35,784	1,60,000
No.of orbits/day	15	2	1	0
Earth's rel motion (/orbit)	24°	−360°	0	−360°
Max scan angle (sat za = 70°)	59.7	13.0	8.2	0.2
Swath (km)	2,293	12,662	13,742	15,508
Obs. Time (hr/pixel)- TIR	0.09	7.6	24.0	9.3
Obs. Time (hr/pixel)- UV/VIS/NIR	0.09	5.7	12.0	9.3
Daily Spatial Coverage	global	global	13,742 km	global
Sun-glint obs	all lats	sub-solar lats	sub-solar lats	sub-solar lats
Multi-angle obs	all lats	tropics	No	tropics
Aperture in cm (200:1 S/N) (5 km pixel, 0.5 sec integ time)	1	36	63	2,837

FIGURE 11. The table shows the tradeoffs in choosing the orbit for AQ studies from satellites. For a nominal measurement requirement (2000:1 S/N at ~400 nm, 5-km pixel, 0.5-sec integration), one requires an instrument of ~1-cm aperture in 564-km orbit. (For reasons discussed in the text, it is difficult to build such an instrument if the required swath is 2293 km, which is needed to get global coverage.) As one moves farther away from Earth to 20,187 km MEO, one continues to get global coverage with increasing observation times per pixel. (Note that the obs time given for UV/VIS/NIR instruments in MEO are mean values; some pixels will get substantially more and some substantially less observation time.) Going to GEO increases the observation time further but reduces the coverage. Going to L1 provides a modest increase in observation time (over MEO), but requires unacceptably large apertures.

While the geostationary orbits are ideal for AQ studies, for they keep an area perpetually in the field-of-view, they provide measurements over less than one-fourth of the Earth's surface area. By moving to lower orbits one allows the orbit to drift over the entire globe but reduces the temporal coverage. An orbit location that some people find attractive is the 1st Langrange point (L1), which lies at ~1.6 million km from the Earth along the Earth–Sun line. For instruments that work with reflected sunlight this location provides full global coverage in one day, with each sunlit pixel remaining in view for close to 9 hours as the Earth revolves under the satellite. However, as the table shows, one requires unacceptably large aperture from this vantage point to study AQ, perhaps with the exception of aerosols, which require a more modest S/N than the others.

Summary

The AQ is controlled primarily by the concentrations of particulates smaller than 2.5 micron and of 4 trace gases that include O_3, SO_2, CO, and NO_2. Satellite instruments, primarily operating from low orbits (< 1000 km altitude), have demonstrated that all five constituents can be measured from space. However, these measurements lack the spatial and temporal resolution deemed necessary for AQ studies. The high vantage point orbits (> 20,000-km altitude) are highly attractive for AQ studies, for instruments in these orbits can provide good temporal and spatial resolution with improved S/N than the LEO instruments. Maximum temporal coverage is provided by instruments in the Geostationary (GEO) orbit, but they can observe less than one-fourth of the globe, thus requiring multiple satellites for global coverage.

UV/VIS/NIR instruments, which use reflected sunlight to make measurements, can use cloudy pixels to sense the atmosphere above the clouds, and cloudfree pixels to sense the atmosphere below the clouds, thus providing some altitude information. Thermal IR instruments have the advantage that they can operate 24 hours a day, but they are sensitive to species in the middle troposphere only, and require cloudfree pixels, which severely limits the useful coverage they can provide from LEO. The high vantage points orbits can greatly improve the number of cloudfree observations one can make, improving the utility of these instruments for AQ studies. Combination of reflected-sunlight and thermal emission instruments would provide highly synergistic measurements of AQ from space.

References

Carn et al. (2003), Volcanic eruption detection by the Total Ozone Mapping Spectrometer (TOMS) instruments: A 22-year record of sulphur dioxide and ash emissions, Oppenheimer, C., Pyle, D.M. & Barclay, J. (eds) *Volcanic Degassing*. Geological Society, London, Special Publications, **213**, 177–202. 0305-8719/03.

Carn S.A., A.J. Krueger, N.A. Krotkov, and M.A. Gray (2004), Fire at Iraqi sulfur dioxide plant emits SO_2 clouds detected by Earth Probe TOMS. *Geophys. Res. Lett.*, **31**, L19105, doi:10.1029/2004GL020719.

Fishman, J., C.E., Watson, J.C. Larsen, and J.A. Logan (1990), Distribution of tropospheric ozone determined from satellite data, *J. Geophys. Res.*, **95**, 3599–3617.

Herman, J.R., P.K. Bhartia, O. Torres, C. Hsu, C. Seftor and E. Celarier [1997], Global distribution of UV-absorbing aerosols from Nimbus7/TOMS data, *J. Geophys. Res.*, **102**, 16911–16922.

Hsu, N.C., J.R. Herman, P.K. Bhartia, C.J. Seftor, O. Torres, A.M. Thompson, J.F. Gleason, T.F. Eck, and B.N. Holben (1996), Detection of biomass burning smoke from TOMS measurements, *Geophys. Res. Lett.*, **23**, 745–748.

Hsu, N.C., S. Tsay, M.D. King, and J.R. Herman (2004), Aerosol properties over bright-reflecting source regions, *IEEE Trans. Geosci. Remote Sens.*, **42**, 557–569.

Kley, D. et al. (1996), Observations of near-zero ozone concentrations over the convective Pacific: Effects on air chemistry, *Science*, **274**, 230–232.

McPeters, R.D., et al. (1996), Nimbu&stilde;7 Total Ozone Mapping Spectrometer (TOMS) Data Product's User's Guide, *NASA Reference Publication 1384,* National Aeronautics and Space Administration, Washington, DC.

Remer, L.A. et al. (2004), The MODIS aerosol algorithm, products and validation, to be published in *J. of Atmos Sci, Special issue: Chesapeake Lighthouse and Aircraft Measurements for Satellites (CLAMS) Field Experiment.*

Richter, A. and J. Burrows (2002), Retrieval of tropospheric NO_2 from GOME measurements, *Adv. Space Res.*, **29**(11), 1673–1683.

Torres, O., P.K. Bhartia, J.R. Herman, Z. Ahmad, and J. Gleason (1998), Derivation of aerosol properties from satellite measurements of backscattered ultraviolet radiation: Theoretical Basis, *J. Geophys. Res.*, **103**, 17,099–17,110.

Wenig et al. (2004), Retrieval and analysis of stratospheric NO_2 from the global ozone monitoring experiment, *J. Geophys. Res.*, **109**, D04315, doi:10.1029/2003JD003652.

Ziemke, J.R., S. Chandra, and P.K. Bhartia (1998), Two new methods for deriving tropospheric column ozone from TOMS measurements: The assimilated UARS MLS/HALOE and convective-cloud differential techniques, *J. Geophys. Res.*, **103**, 22,115–22,127.

Ziemke, J.R., S. Chandra, and P.K. Bhartia (2001), "Cloud slicing": A new technique to derive upper stratospheric ozone from satellite measurements, *J. Geophys. Res.*, **106**, 9853–9867.

Chapter 2

Aerosol Forcing and the A-Train

CHIP TREPTE

Introduction

Aerosols are small particles consisting of solid or liquid material suspended in air, with diameters ranging from about 0.01 to 1 μm. They originate primarily from natural sources such as volcanoes, dust storms, forest fires, vegetation, and sea spray. Anthropogenic sources include burning of fossil fuels and land use changes such as biomass burning and deforestation. Depending upon their size, injection altitude, and proximity to precipitating cloud systems, lifetimes of aerosols in the atmosphere vary considerably from a few hours to as much as 2 weeks and thus, can be transported long distances from their sources.

Aerosols can have a significant impact on the environment and climate. They can have a *direct* radiative forcing effect by scattering and absorbing sunlight. The change in direct radiative forcing due to anthropogenic aerosols alone (also known as direct aerosol climate forcing) is substantial and rivals that due to stable greenhouse gases. For the most part it is opposite in sign to the greenhouse gases, however, the regional nature of aerosol forcing and its variable vertical distribution precludes simple conclusions of a global canceling of greenhouse gas warming.

Aerosols can also *indirectly* influence radiative forcing by modifying the albedo of clouds either through the incorporation of absorbing aerosols, which tend to darken them, or through the seeding of smaller droplets, which tend to brighten them. It has been suggested that aerosols can alter the liquid water content of clouds, thereby lengthening cloud lifetime and the geographical extent of cloudiness, or even change the vertical structure of latent heating and influence atmospheric dynamics.

At present, our ability to quantify and predict the impact of global aerosol forcing on climate remains highly uncertain [IPCC, 2001]. The uncertainties arise, in part, because of insufficient knowledge of aerosol variability, composition, optical properties, hygroscopicity, and size distribution. Complex aerosol processes such as transport, transformation, aging, and cloud interactions are also poorly represented in models. More observations from a variety of techniques, platforms, and vantage points are needed to improve our characterization of the aerosol system and reduce modeling uncertainties.

In the next 12 months, the A-train satellite constellation consisting of the Aqua, CloudSat, CALIPSO, PARASOL, and Aura satellite missions will be fully assembled in orbit and will provide valuable new information to help address some of these uncertainties. For the first time, a three-dimensional view of the aerosol and cloud system over the globe will be available from a combined instrument suite including a lidar, a radar, multispectral imagers and broadband radiometers. These measurements will enable more accurate estimates of direct aerosol forcing than presently available. They will be made above, below, and next to cloud systems; over bright and dark surfaces; and at high and low latitudes. The measurement suite will further permit a better understanding of regional impacts of indirect aerosol forcing and the evolution of aerosol distributions that will help to constrain parameterizations of aerosol processes in climate models.

This lecture will provide an overview of the basic factors governing direct and indirect aerosol radiative forcing. It will be followed with a summary of the measurement capabilities for aerosols on the A-Train constellation and on how these might be utilized to better understand aerosol processes and provide more accurate estimates of aerosol forcing.

Direct Aerosol Forcing

In the early 1990s it was believed that radiative forcing by tropospheric aerosols was dominated by sulfate droplets such as those found in the stratosphere following large volcanic eruptions. These aerosols absorbed negligible amounts of visible light and, hence, have a single scattering albedo (fraction of light attenuated by scattering to the total amount attenuated by scattering and absorption) close to unity [Charlson et al., 1992]. Over the years, however, greater realization of the importance of different aerosol types and their optical properties revealed that aerosol forcing is more complicated and may, under suitable conditions, cause net heating.

The following expression [Haywood and Shine, 1995] is used to identify how different factors influence aerosol radiative forcing:

$$\Delta F = -DS_o T_{at}^2 (1 - A_c)\omega\beta\delta\left((1 - R_s)^2 - \frac{2R_s}{\beta}\left(\frac{1}{\omega} - 1\right)\right)$$

where D is daylight, S_o is the solar constant, T_{at} is the atmospheric transmission above the aerosol layer, A_c is fractional cloud cover, R_s is the surface reflectivity, ω is the single scattering albedo (fraction of light attenuated by scattering to the total amount attenuated by both absorption and scattering), δ is optical depth, and the β is the upscatter fraction. The expression reveals that in clear skies ΔF is governed principally by aerosol abundance through δ, aerosol optical properties through β and ω, and the underlying surface reflectance.

Further inspection of the above equation reveals that forcing can shift from net cooling to net warming when

$$\omega < \frac{2R_s}{\beta(1 - R_s)^2 + 2R_s}$$

Typical values of the upscatter fraction β range from .15 to .30. For sulfate aerosols (ω ~ 1), the radiative forcing is negative over almost all surface conditions. However, for aerosols that absorb solar radiation (e.g., containing black carbon) the single scattering coefficient decreases, net warming occurs when the particles overlie gray or bright surfaces. For example, when $\omega = 0.9$ and $R_s > 0.3$, the atmosphere will warm. This example is simplified to illustrate a few of the complexities involved in modeling direct aerosol forcing. More accurate representation would include factors with possible interdependencies such as hygroscopicity, internal/external mixtures, varying solar angles, etc.

Since climate forcing is due only to the anthropogenic component of the aerosol distribution, we need isolate this contribution from the total radiative forcing calculation. This aspect is best performed by chemical analysis with in situ instrument probes. An upper estimate of global climate forcing from satellite observations may be obtained by realizing that anthropogenic aerosols often fall within the fine mode fraction as suggested by a conceptual representation of the aerosol size distribution. Coarse mode particles are usually formed by mechanical actions (e.g., wind-blown dust or sea salt). Fine mode particles often form as a result of combustion or nucleation/condensation. The difficulties of isolating the anthropogenic component and the need for global observations underlines the need for an integrated measurement sensor web that includes ground-base, aircraft, satellite observations fused together by data assimilation techniques [Diner et al., 2004; Anderson et al., 2004].

Indirect Aerosol Forcing

Since aerosols are the building blocks of clouds, it is natural to expect that changes in aerosol characteristics can impact the properties of clouds. Twomey [1977] first described how the presence of more cloud condensation nuclei could increase the liquid water content (LWC) of clouds and change their cloud optical properties.

To illustrate the relationship, consider the definition of optical depth,

$$\delta = \pi h \int_0^\infty Q_e r^2 n(r) dr$$

where h is the depth of the cloud, Q_e is the extinction efficiency, r is the particle radius, and $n(r)$ is the size distribution. At visible wavelengths $Q_e \sim 2$. If we further consider a narrow droplet-size spectrum with a mean radius, $\bar{r}$, the above expression can be approximated as

$$\delta = 2\pi h \bar{r}^2 N$$

where

$$N = \int_0^\infty n(r) dr$$

is the total number concentration of droplets. The total liquid water content of a cloud is defined as

$$LWC = \frac{4\pi}{3}\rho \int_0^\infty r^3 n(r)dr$$

After making similar substitutions for the droplet size distribution and mean radius the expression can be approximated by

$$LWC = \frac{4\pi}{3}\rho \bar{r}^3 N$$

Relating together the approximated expressions for optical depth and liquid water content yields

$$\delta = 2.4\left(\frac{LWC}{\rho}\right)^{2/3} hN$$

This has the consequence that an increase in cloud condensation nuclei produces an increase in optical thickness. Because of the increased competition for water vapor with the increased aerosol population, cloud droplets may be unable to grow large enough to initiate the collision-coalescence process (the dominate precipitation process for warm clouds). The prevailing thought is that precipitation rates, thus, will decrease and cloud lifetimes lengthen.

A number of studies have observed changes in aerosol concentrations, cloud droplet number/size, and cloud reflectance. Many of these studies include both aircraft and satellite observations at near-infrared wavelengths, where cloud reflectance is highly sensitive to droplet size rather than optical thickness [e.g., Nakajima et al., 2001]. Some of the most compelling evidence showing changes in aerosol optical properties by aerosol concentration comes from AVHRR observations of the effects of ship exhaust on clouds [Coakley et al., 1987]. Because of the magnitude and degree of uncertainty with indirect aerosol forcing, more observations and more modeling research is needed. Further investigations from the advantage point of space, hopefully, can continue to provide an improved understanding of the regional and global affects of aerosol-cloud interactions and their impact on climate.

A-Train Aerosol Measurements

The A-Train satellite constellation (named for the Aqua and Aura satellites) has an orbit at an altitude of 705 km and inclination of ~98°. Aqua leads the constellation with an equatorial crossing time of approximately 1:30 p.m. and Aura trails the set with a crossing time of ~1:45 p.m. On December 2004, the PARASOL satellite was launched followed by the joint launch of CALIPSO and CloudSat on April 2006. These satellites have been joined the constellation in orbit between the Aqua and Aura platforms. Table 1 summarizes the instruments on the different platforms and products available for aerosol studies. As seen in the table, a large number of near-coincident measurements on aerosol and cloud physical and optical properties, radiative flux, and thermodynamic parameters will be available over the same locations within a few minutes of each other. (Details on the

A-train constellation may be found at the following Web site: http://eospso.gsfc.nasa.gov/eos_homepage/for_educators/educational_publications.php.)

Global maps of aerosol optical depth can be derived from measurements by MODIS, PARASOL and OMI. These instruments exploit the spectral sensitivity of the size distribution and composition of aerosols to infer their optical properties. For MODIS, visible and near-IR measurements over dark surfaces are used to retrieve aerosol optical depth [Kaufman et al., 2002] and provide discrimination between accumulation and coarse size modes [Remer et al., 2002]. Over bright surfaces, less spectral information is available because of the lack of spectral contrast and the complications of large variability in surface bidirectional reflectance. An example map of aerosol optical depth derived from MODIS is shown in Figure 1 for the months of January and July 2003. For the month of January, most oceanic regions have $\delta < 0.2$. However, near the western coast of Africa and over central Africa values peak around 0.8. High values are also present over eastern China. For the month of July, much higher aerosol loading is present over the globe with peaks in regions near or downwind of the major deserts and large areas of industrial activity. The white regions in both figures are data void areas (note that they occur over high albedo regions). It should be noted that application of the technique is permitted only for cloud free pixels. The data shown represents the cumulative average for available scenes for the month. These were obtained from a catalogue of aerosol data products from the MODIS Online Visualization and Analysis System at (http://lake.nascom.nasa.gov/www/ online_analysis/movas).

Aerosol retrievals using POLDER observations on PARASOL are expected to follow the technique demonstrated Bellouin et al. [2003]. The approach uses multiangle images at 9 different wavelengths (443, 490, 565, 670, 763, 765, 865, 910, and 1020 nm) with channels at 490, 670, and 865 nm sensitive to polarized signals to derive optical depth and information on aerosol size distribution (Angstrom coefficient) over the ocean. Over land, the technique provides an index that is proportional to the product of optical depth and the Angstrom coefficient. The OMI aerosol retrieval uses near-UV spectral channels instead of those used by MODIS in the mid-visible region. The advantage in this approach is that most land surfaces are dark in the near UV and the interaction between aerosol and Raleigh scattering offers sensitivity to aerosol absorption [Torres et al., 2002]. This feature permits OMI to provide insight on the bulk distribution of single scattering albedo over the globe.

The addition of CALIPSO mission to the A-train constellation greatly complements the other aerosol observations by providing information on the vertical structure of aerosol and cloud layers [Winker et al., 2003]. CALIPSO consists of a polarization-sensitive lidar operating at 1064 and 532 nm and passive imagers operating in the visible and infrared spectral regions. It has a designed optical depth sensitivity of at least 0.005 at 532 nm. With this information, aerosol features can be unambiguously identified above and between clouds and over bright surfaces. The technique is especially sensitive at low optical depths and promises

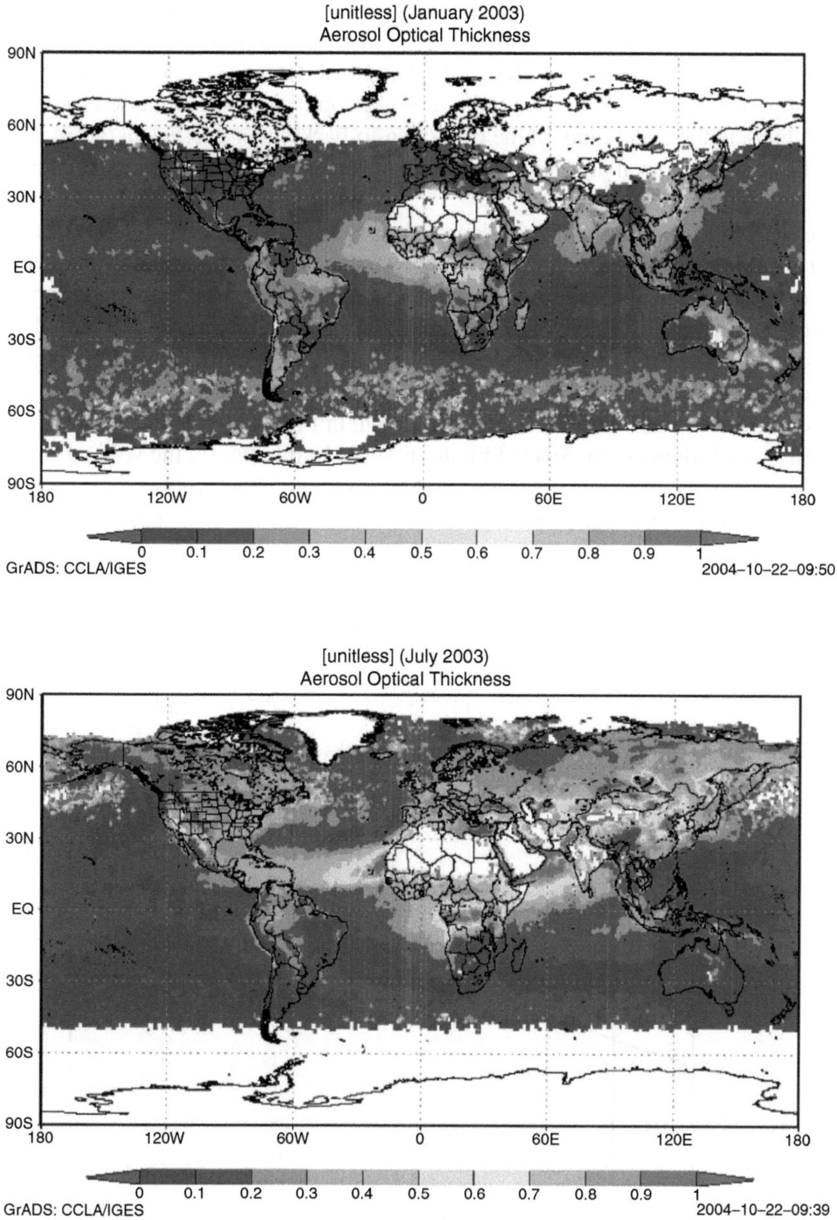

FIGURE 1. Aerosol optical depth obtained from the MODIS satellite experiment for the months of January and July 2003.

to improve our quantitative estimate of global aerosol forcing. This is illustrated in Figure 2, which shows the estimated uncertainty derived from CALIPSO compared with retrievals from MODIS. At low optical depths (< ~.2), lidar retrievals are superior. Their accuracy degrades at higher optical depths because of greater weighting in the uncertainty of the extinction-to-backscatter coefficient that is used in the aerosol retrieval algorithm. This parameter is dependent on composition and the size distribution.

CALIPSO's measurements of aerosol layers will also be valuable in evaluating model transport processes. It is widely recognized that chemical-transport models still have difficulty in reproducing observed vertical distributions of aerosols. The ability to distinguish aerosol layers within or above the boundary layer will help to identify transport pathways and provide insight on their region of origin with the use of backtrajectory analysis and tracer observations. An example of a set of lidar observations is shown in Figure 3 from the LITE experiment, which flew on the Space Shuttle in 1994. In this figure, the vertical structure of clouds and aerosols are clearly evident. The yellow lines show backtrajectories and highlight the nature of differential transport processes in the atmosphere.

Another area where the CALIPSO data will provide new information is in the use of the ratio of 532/1064-nm backscatter measurements. According to Mie

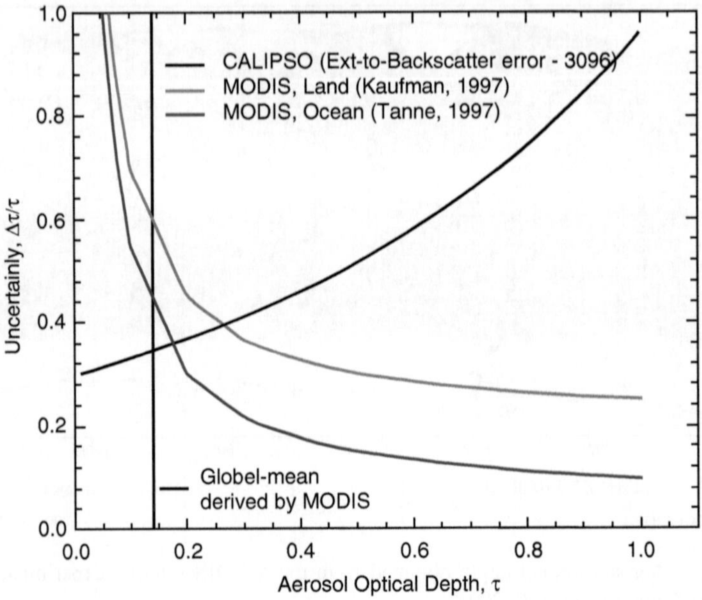

FIGURE 2. Uncertainty estimate for retrieved aerosol optical depth. The CALIPSO algorithm assumes an uncertainty of 30% in the extinction-to-backscatter coefficient. The two different error estimates from MODIS reflect the use of different retrieval techniques for either land or oceanic surfaces.

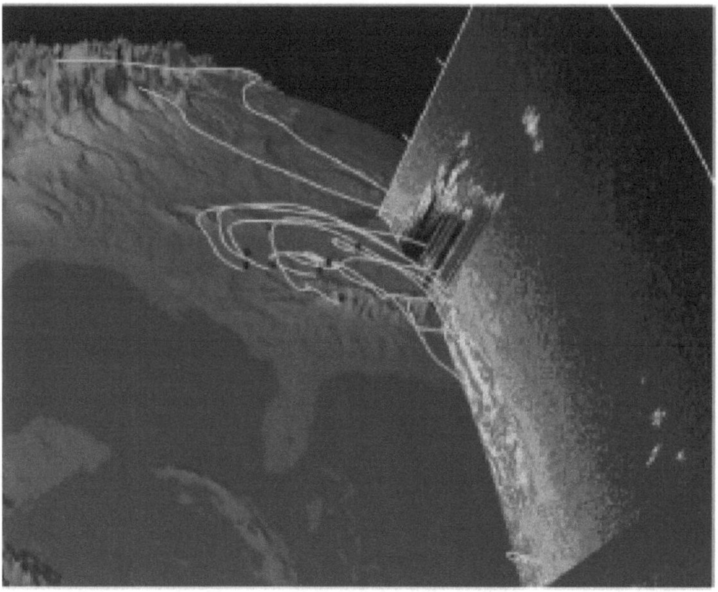

FIGURE 3. Example cross-section of lidar backscatter observations obtained from the LITE experiment over North America. Aerosol layers are denoted by warm colors. Clouds are indicated by white features. The yellow lines are backtrajectories.

theory, ratios approaching unity suggest the presence of larger particles, while those at lower values suggest the presence of smaller particles. This information coupled with absolute backscatter or extinction values can point to the likelihood of cloud, coarse, or fine mode particles. From this data, clouds and aerosols can be separated with confidence as a function of altitude. This capability may prove valuable in tuning of cloud mask algorithms needed for aerosol and cloud retrievals from MODIS and POLDER.

For understanding interactions between aerosols and clouds, CloudSat will provide key information on the properties of optically thick clouds, their ice and liquid water content, and associated rates of precipitation with its 94-GHz radar [Stephens et al., 2002]. Reflectivity measurements will be obtained from the surface to 30 km with 500-m vertical resolution. Because cloud properties can change rapidly, the CloudSat platform will be placed in orbit ~2 minutes behind Aqua, with CALIPSO trailing CloudSat by ~15 sec. In addition, control of CloudSat's cross-track motion will be within ±1 km of the CALIPSO ground track to ensure the most lidar/radar observations are co-aligned.

Concurrent observations of thermodynamic state parameters (e.g., temperature, relative humidity) from instruments on Aqua will provide the ability to examine interrelationships between aerosol optical depth, single scattering albedo, and relative humidity. Importantly, concurrent A-train aerosol observations with radiative flux measurements from CERES will enable more accurate

TABLE 1. A-Train Instrument Characteristics and Related Products

Spacecraft	Instruments	Characteristics	Aerosol/Cloud Products
Aqua	MODIS	36-channel visible radiometer 2300-km swath Variable pixel resolution .25 to 1 km	Land, ocean, and atmospheric products; include cloud and aerosol optical depths and particle size information
	AIRS/AMSU-A	IR and microwave sounders, 1650-km swath IR pixel resolution ~10 km	Temperature and moisture profiles in clear atmosphere
	AMSR-E	6-channel microwave radiometer 1445-km swath	Liquid water path, column water vapor, liquid precipitation
	CERES	Broadband and spectral radiances converted to fluxes Pixel resolution ~20 km	Top of atmosphere radiation budget Time mean and instantaneous fluxes
CloudSat	94-GHz radar	500-m vertical range gates from surface to 30 km Pixel resolution ~1.4 km	Cloud profile information, liquid and ice water content, precipitation
CALIPSO	Lidar	532- and 1064-nm channels with depolarization Field of view 70 m at surface with 333-m separation between samples	Aerosol and cloud profile for optically thin clouds Aerosol extinction, optical depth, and typing
	IIR/WFC	3-channel IR radiometer Visible wide field of view camera 64-km swath IR pixel resolution 1 km WFC pixel resolution 125 m within 2.5 km of nadir and 1 km beyond	Cirrus cloud properties
PARASOL	POLDER	9-channel polarimeter with channels in visible and near IR. 400-km swath Pixel resolution 5 km	Cloud and fine mode aerosol optical depth and particle sizes
Aura	HIRDLS	IR limb sounder	Trace gases and stratospheric aerosol
	MLS	Microwave limb sounder	Trace gases, ice content of upper tropospheric cloud
	TES	IR imaging spectrometer Pixel resolution .5 × 5 km	Trace gases
	OMI	UV grating spectrometer Pixel resolution 13 × 24 km	Ozone, aerosol optical depth and single scattering albedo

observationally based estimates of aerosol forcing over the globe. This method has already been demonstrated by Haywood et al. [1999] and Christopher and Zhang [2002] using MODIS/CERES data. The added information from CALIPSO and CloudSat should improve estimates near and in cloudy regions and regions with bright surfaces.

Acknowledgements Some of the data used in this lecture were acquired as part of the NASA's Earth Science Enterprise. The MODIS aerosol retrievals were provided by MODIS Science Teams and data processed by the MODIS Adaptive Processing System (MODAPS). These are archived and distributed by and Goddard Distributed Active Archive Center (DAAC).

References

Bellouin, N., O. Boucer, D. Tanre and O. Dubovik, Aerosol absorption over the clear-sky oceans deduced from POLDER-1 and AERONET observations, Geophys. Res. Lett., **30**(14), 1748, doi:10.1029/2003GL017121, 2003.

Charlson, R.J., S.E. Schwartz, J.M. Hales, R.D. Cess, J.A. Coakley, J.E. Hansen, and D.F. Hofmann, Climate forcing by anthropogenic aerosols, Science, **255**, 423–430, 1992.

Coakley, J.A., R.L. Bernstein, and P.A. Durkee, Effect of shipstack effluents on cloud reflectivity, Science, **237**, 1020–1022, 1987.

Husar, R., J.M. Prospero, and L.L. Stowe, Characterization of tropospheric aerosols over the oceans with the NOAA advanced very high resolution radiometer optical thickness operational product, J. Geophys. Res., **102**, 16889–16909, 1997.

Intergovernmental Panel on Climate Change, Aerosols, their direct and indirect effects, ed., J. Penner et al., Climate Change 2001, The Sciencitific Basis. Working Group I to the Third Assessment Report of the IPCC, Camridge University Press, 2001.

Kaufman, Y.J., and T. Nakajima, Effect of Amazon smoke on cloud microphysics and albedo – Analysis from satellite imagery, J. Appl. Meteor., **32**, 729–744, 1993.

Nakajima, T., A. Kawamoto, and J.E. Penner, A possible correlation between satellite-derived cloud and aerosol microphysical parameters, Geophys. Res. Lett., 2001.

Remer, L.A., D. Tanre, Y.J. Kaufman, C. Ichoku, S. Mattoo, R. Levy, D.A. Chu, B. Holben, O. Dubovik, A Smirnov, J. Martins, R. Li, and Z. Ahman, Validation of MODIS aerosol retrieval over ocean, Geophys. Res. Lett., **29**, 10.1029/2001GL013204, 2002.

Stephens, G.L., D.G., Vane, R.J. Boain, G.G. Mace, K. Sassen, Z. Wang, A. Illingworth, E O'Conner, W. Rossow, S. Durden, S. Miller, R. Austin, A. Benedetti, C. Mitrescu, The CloudSat mission and the a-train; a new dimension of space-based observations of clouds and precipitation, Bull. Amer. Meteor., **83**, 12, 1771–1762, 2002.

Torres, O., R. Decae, and P. Veefkind, OMI Aerosol Retrieval Algorithm. In OMI Algorithms Theoretical Basis Document, Volume III, Clouds, Aersols, and Surface UV Irradiance, P. Stammes (ed.), NASA-KNMI, 2002.

Twomey, S.A., The influence of pollution on the shortware albedo of clouds, J. Atmos. Sci., **34**, 1114–1152, 1977.

Winker, D.M., J. Pelon, and M.P. McCormick, The CALIPSO mission: Spaceborne lidar for observation of aerosols and clouds, Proc. SPIE, **4893**, 1–11, 2003.

Chapter 3

Total Ozone from Backscattered Ultraviolet Measurements

Pawan K. Bhartia

Introduction

The outline of this lecture is as follows. First, we look at how total O_3 is derived by looking at the sun from ground-based instruments. This technique was developed by Prof. Dobson in the UK more than half a century ago; still considered the "gold standard" against which all other techniques—both satellite and ground-based—are judged. We discuss several ways of retrieving total O_3 from direct sun measurements. Understanding the pros and cons of these methods help us understand the differences between the various methods of producing total O_3 from satellite. Finally, we discuss the TOMS total O_3 algorithm and show key results produced from this instrument from a record that now spans more than a quarter century.

Total Ozone from the Measurement of Transmitted Solar Flux

Referring to Figure 1, the basic quantity one attempts to measure is the atmospheric transmittance $T(\lambda)$, at wavelength λ, which is defined as the ratio of solar irradiance at the ground $F(\lambda)$, and the solar irradiance at the top of the atmosphere, $F_0(\lambda)$. From the Lambert–Beer law, $T(\lambda)$ can be expressed in terms of $T_0(\lambda)$, the transmittance in absence of ozone absorption, multiplied by an exponential term that represents the ozone absorption. The variable m in the exponent is called the air mass factor (AMF), which can be written approximately as $\sec\theta_0$ (by ignoring the sphericity of the Earth's atmosphere), where θ_0 is the solar zenith angle (SZA). The second term in the exponent is the O_3 optical depth (τ), which can be written as the product of the ozone absorption cross-section (σ_{O3}) and total number of O_3 molecules per square area, N_{O3}. σ is usually expressed in cm^2, and N in molecules/cm^2. However, for total ozone it is customary to use the Dobson Unit (1 DU = 2.67 10^{16} molecules/cm^2), which is named after the British scientist who first measured ozone using the

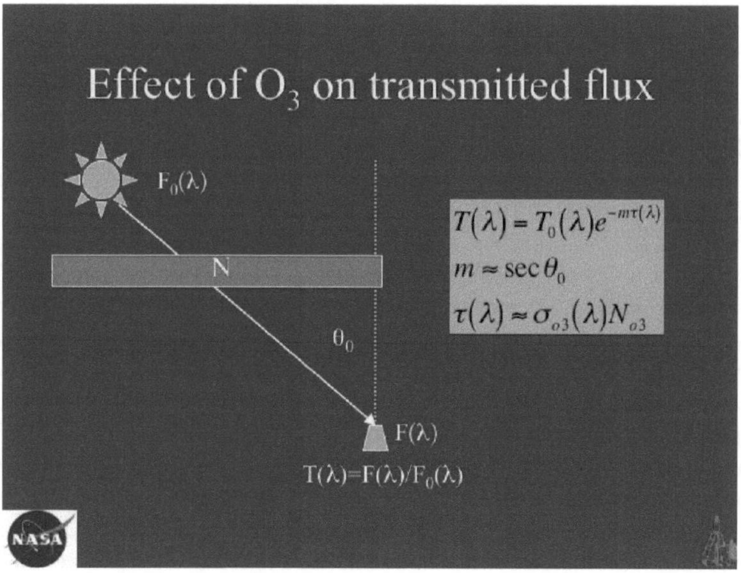

FIGURE 1. In the direct-sun technique one measures the solar radiation transmitted through the atmosphere, from which the total number of O_3 molecules per unit area, N, can be derived using the formula shown.

direct-sun technique [Dobson, 1931]. The amount of ozone in the Earth's atmosphere

However, the large cubic polynomial term in the ozone cross-section complicates the analysis by varying from less than 100 DU inside the Antarctica polar vortex (O_3 hole) and can occasionally reach 600 DU in the Feb/March months in the northern hemisphere. Typical values are around 300 DU.

In Figure 1, the red band shows a simplified representation of the O_3 layer. It is assumed that the O_3 molecules are concentrated in a thin layer and are distributed homogeneously throughout the layer. The Lambert–Beer formula, as written, is valid only under this assumption. In the Earth's atmosphere, ozone is distributed inhomogeneously over a broad range of altitudes. We will consider the impact of this later in this lecture.

To estimate N_{O3}, we need a method of estimating T_0. At ultraviolet wavelengths where O_3 absorption is weak (340–400 nm) Rayleigh scattering by the molecular atmosphere is the principal source of attenuation of the solar radiation, followed by aerosol attenuation, both of which cause T_0 to vary smoothly with wavelength. Therefore, if there are no other absorbers besides O_3, one can assume that this behavior continues at shorter wavelengths at which O_3 is measured (300–340 nm) from solar data. This is the basic assumption made in the Differential Optical Absorption Spectroscopy (DOAS) technique (Figure 2), in which one assumes that the variation of $\ln T_0$ with λ can be accurately represented by a 3rd-order polynomial.

As shown in the attached figure, with this assumption, one can, in principle, estimate N by linear least-squares regression. However, if one examines the spectral dependence of ozone absorption cross-section (Figure 2 inset graph) one finds that it contains a large 3rd-order polynomial term of its own. So, in the actual fit one estimates N from the residual absorption $\Delta\sigma(\lambda)$, rather than the total absorption $\sigma(\lambda)$. Note that the wiggles in $\Delta\sigma(\lambda)$ are smaller than 10^{-20}; so, for typical values of $N\sec\theta_0$ (~10^{19} molecules/cm^2), one must fit the spectrum to few parts in 10^4 to achieve 1% accuracy in measuring N. It is very difficult to measure $T(\lambda)$ to such accuracy even with modern instruments, and was impossible to do so 70 years ago when such measurements got started. However, if one can somehow take advantage of the much larger total absorption, then instrument accuracy/precision requirements can be greatly simplified. The most general method of doing this is to do a two-step retrieval. In the first step, the coefficients of the 3rd-order polynomial representing $\ln T_0(\lambda)$ can be derived from measurements taken at wavelengths where the O_3 absorption is negligible (340–400 nm); then, in the second step, these coefficients can be used to remove the polynomial from $\ln T$ prior to fitting the measurements in the 310–340 nm wavelength range, thus allowing one to estimate N_{O3} using the total absorption. Although the exact

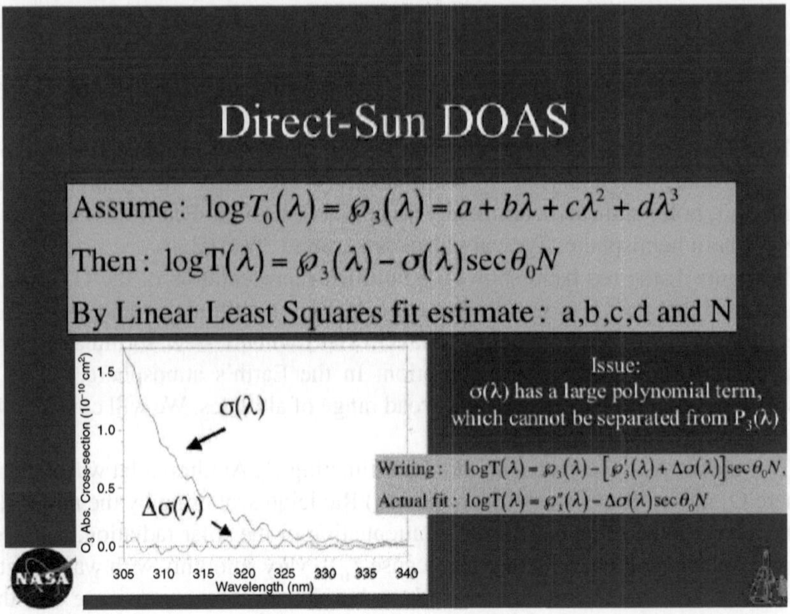

FIGURE 2. In the DOAS technique, one estimates N by linear least-squares regression assuming that the attenuation of the solar radiation in absence of ozone can be expressed by a cubic polynomial.

method just described hasn't been widely used so far, the technique used to estimate N_{O3} from Dobson and Brewer spectrophotometer instruments uses a simplified version of this concept.

In the Dobson–Brewer technique, one first reduces the order of the $\ln T_0$ polynomial by explicitly accounting for the strong attenuation of the solar ultraviolet radiation by Rayleigh scattering. The Rayleigh optical depth τ_R (~1 at 310 nm), is proportional to the surface pressure and is known to better than 1%, so it is easy to estimate [Bates, 1984]. The remaining component of T_0 is Mie scattering by aerosols T_m. By assuming that $\ln T_m$ varies linearly with wavelength, one can derive the coefficients of the linear polynomial using a pair of wavelengths at which the O_3 absorption is weak; N_{O3} can then be estimated using a 3rd wavelength. However, for historical reasons, the Dobson–Brewer technique uses two wavelength pairs. Figure 3 shows graphically how the technique works. One notes that the differential absorption one uses in this technique is about 30 times larger than the differential absorption of the DOAS technique; requiring ~1% measurement accuracy to measure N_{O3} to 1%, greatly simplifying the instrument design and measurement accuracy/precision requirements.

Before we end this section, we briefly discuss the impact of the approximations made in Figure 1. As shown in Figure 4, the curvature of the Earth's atmosphere causes the airmass factor of an atmospheric layer to change with altitude, and the

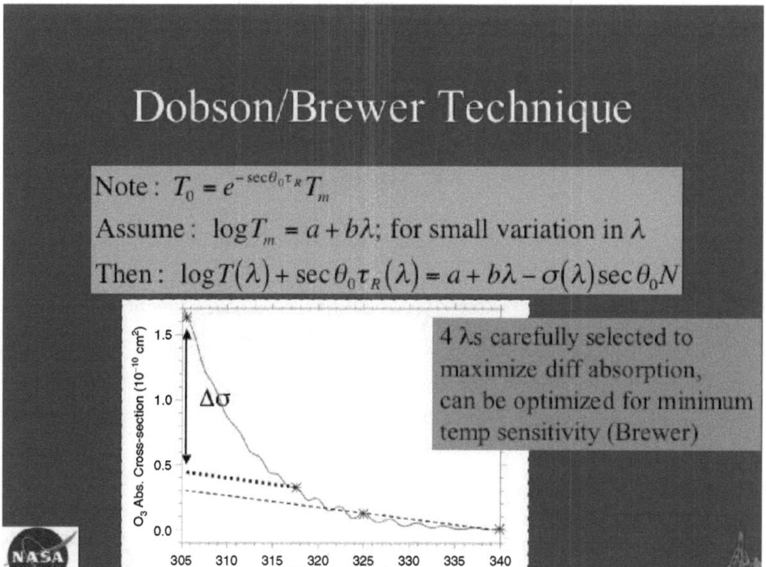

FIGURE 3. In the Dobson–Brewer technique, one estimates N_{O3} by using two pairs of wavelengths; the longer wavelength pair is sensitive primarily to the differential Mie scattering by aerosols, the shorter wavelength pair to ozone.

temperature dependence of the O_3 cross-section causes the O_3 absorption optical depth of a layer to do the same.

However, one can preserve the Lambert–Beer formula by substituting the average values of m and σ_{O3} in the formula, as shown in Figure 4. To calculate these averages one must have some *a priori* knowledge of the shape the O_3 profile. Since most of the O_3 in the Earth's atmosphere is located between 15–35 km, the *a priori* dependence is rather small, except at very large solar zenith angles (>80°). Although, for instrumental reasons, Dobson–Brewer measurements are rarely taken above 70° SZA, other ground-based techniques (using zenith sky radiances) can go up to 90° SZA, and direct-sun measurements from high altitude aircraft have been taken up to 92° SZA. Retrieval of N_{O3} from these measurements requires good *a priori* knowledge of the shape of the O_3 profile.

This discussion is designed to highlight the fact that almost all remote sensing techniques require some sort of *a priori* information to do retrieval. Consequently, the accuracy and precision of the retrieved quantities is determined not just by the accuracy and precision of the basic measurements, but also of the *a priori* information. This is an important difference between laboratory and *in situ* techniques and remote sensing techniques. When one studies long-term trends from remote sensing techniques one must consider how long-term changes in the assumed profiles may have impacted the derived trends. It is well known that the shape of the O_3 profile has changed over the last 2 decades due to depletion of the stratospheric O_3 and increase of the tropospheric O_3. This change can affect the trends derived from the remote sensing data and must be considered in the analysis.

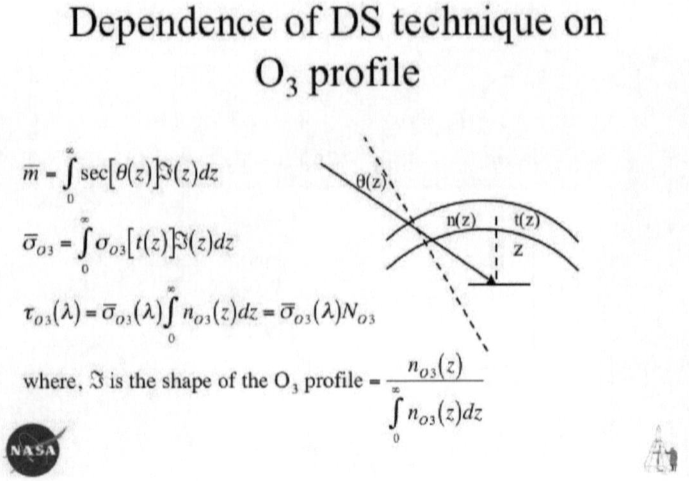

FIGURE 4. The curvature of the Earth's atmosphere and temperature dependence of O_3 absorption cross-section causes m and τ_{O3} to become dependent on O_3 profile. Retrieval of N_{O3} requires *a priori* assumptions about the shape of the O_3 profile.

Total O_3 from Satellite Ultraviolet instruments

The basic measurement one makes from satellites is the reflectance at the top of the atmosphere, ρ. The definition of ρ given in Figure 5 is such that a perfectly reflecting Lambertian sphere would produce a ρ of unity at all solar zenith angles. Since Earth is not a Lambertian sphere, ρ varies with solar zenith angle, as well as with viewing geometry (satellite zenith and azimuth angles). Although, as shown in Figure 5, the scattering, reflection, and absorption processes that determine ρ are quite complicated, it turns out that, with some approximation, one can write a formula for ρ that looks very similar to the one for the direct-sun transmittance (Figure 1). This leads to retrieval techniques that are conceptually similar to the DS retrieval techniques, though, as we shall see, there are many differences in detail.

Figure 6 describes the application of the DOAS technique to derive N_{O3} from the measurement of satellite reflectances. One immediately notes a new term in $\ln\rho_0$. This term is introduced by inelastic scattering by the molecular atmosphere (Rotational Raman scattering), in which the wavelength of the incoming and the scattered radiation changes by a small amount (~1 nm). Despite the fact that the wavelength shift is small and only few percent of the scattering is inelastic, this effect (also called the Ring effect) introduces spectral structures in $\ln\rho_0$ that must be corrected to derive N_{O3} accurately. In the early versions of the DOAS algorithms, a Ring spectrum was assumed and its magnitude was derived by adding a term in the DOAS fit. More advanced algorithms [Valks et al., 2004]

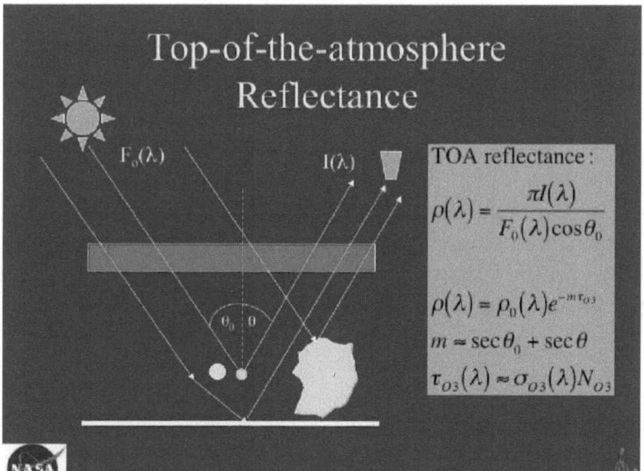

FIGURE 5. In the 310–340-nm wavelength range, top-of-the-atmosphere reflectance measured by satellite is approximately related to N_{O3} by a relationship similar to that for direct-sun transmittance.

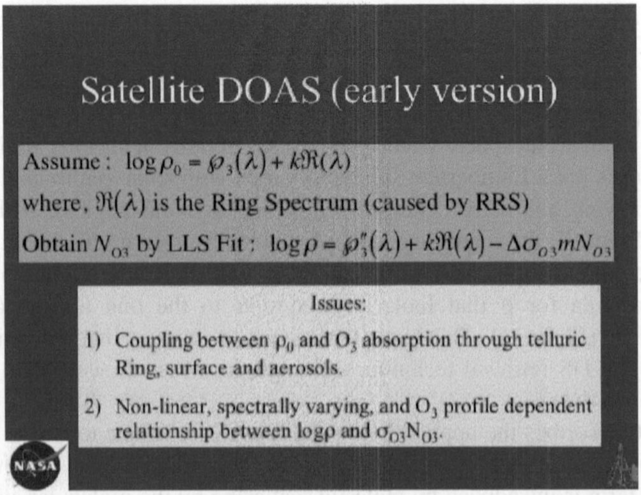

FIGURE 6. Application of the DOAS algorithm to satellite measured reflectances is conceptually similar to that for DS, but requires additional terms, like the Ring effect and correction for some nonlinear effects.

account for the Ring effect caused by the O_3 absorption lines themselves (called telluric Ring) to improve accuracy. These algorithms also account for several other effects, some of which are listed in Figure 6, in an approximate way.

It is worth mentioning two other more advanced algorithm options, which are described in Figure 7. In the advanced DOAS algorithm, listed as Option 1, the nonlinear contributions to the reflectance from molecular and Mie scattering, surface reflectance, Ring effect, etc., are explicitly calculated using a radiative transfer model, and an *a priori* ozone profile. The retrieval algorithm then solves for change in N_{O3} (from the *a priori*) by a linear least-squares fit on the difference between the measured and calculated reflectances. This algorithm works best for wavelengths longer than 320 nm [Coldewey-Egbers et al., 2004, Yang et al., 2004]. In option 2, one does maximum likelihood estimation to derive an O_3 profile using shorter wavelengths, N_{O3} is then just the integrated amount in the retrieved O_3 profile [Bhartia et al., 2004]. The latter algorithm has the advantage that it is less dependent on on *a priori* O_3 profile, particularly at large solar zenith angles.

TOMS Total O_3 Algorithm

The TOMS total O_3 algorithm is a simplified version of the Option 1 algorithm described in Figure 7. The simplification is necessary because like Dobson and Brewer the TOMS instrument has limited number of wavelengths (6), which precludes the application of the full spectral-fitting algorithms described above.

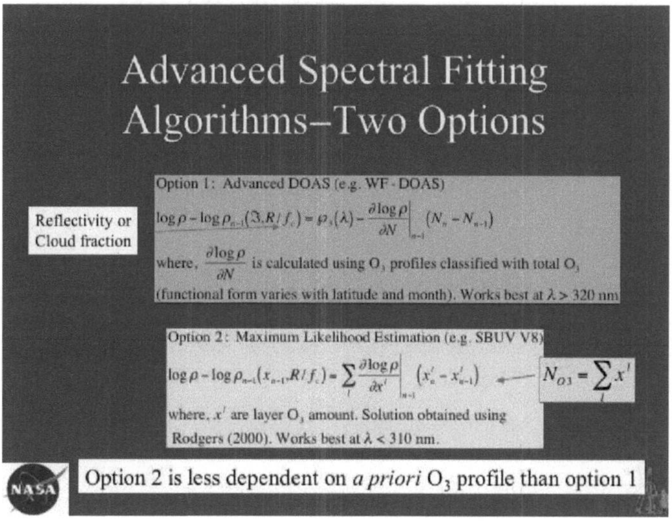

FIGURE 7. Two advanced spectral fitting algorithms recently developed for processing GOME and SBUV data are briefly described.

Yet, with 25 years of operational experience, the TOMS algorithm has become quite mature and has achieved accuracies comparable to the best-run Dobson–Brewer stations. The concepts developed for the TOMS algorithm have now been adopted by the DOAS and the more advanced spectral-fitting algorithms. This section provides an overview of these concepts.

Figure 8 shows the cover of a 39-year-old paper by Dave and Mateer [1967] that first described two key ideas for deriving total ozone from satellite data. With some modifications these ideas were adopted to process early SBUV data [Klenk et al., 1982]. First of these important ideas is the recognition that latitude and total ozone are by far the best predictors of the O_3 profile at any given time and location on the Earth [Wellemeyer et al., 1997]; hence they used total ozone itself to select from a set of *a priori* profiles, starting from a first guess total ozone and then refining it iteratively. Although month has also been added as a predictive variable for profiles in the more recent TOMS algorithm (version 8), as well as in one implementation of the DOAS algorithm [Wellemeyer et al., 2004], concept of using total O_3 as a predictive variable for O_3 profile was and still remains a powerful concept.

Another important idea advanced by the Dave and Mateer paper was to treat Earth's surfaces and clouds as opaque Lambertian reflectors and to derive their reflectivity using measurements at a weak ozone absorbing wavelength. Dave [1978] later called the reflectivity derived using this assumption Lambert-equivalent reflectivity (LER) in recognition of the fact that terrestrial and ocean surfaces are usually non-Lambertian (i.e., their reflectivity varies with solar zenith angle and viewing geometry), and clouds are usually neither opaque nor

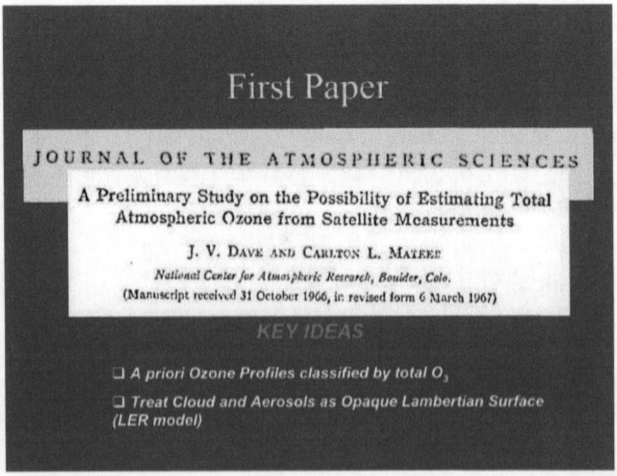

FIGURE 8. First paper describing the technique for measuring total O_3 from satellites. TOMS uses a modified version this technique.

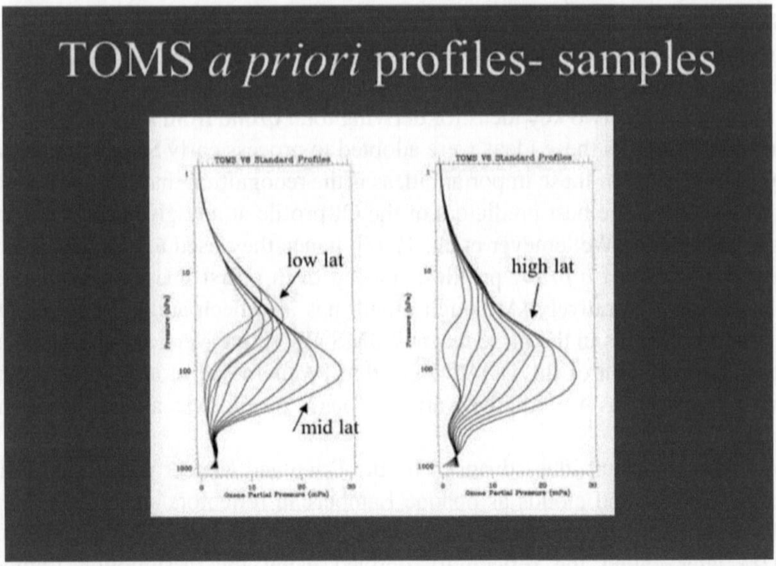

FIGURE 9. TOMS V8 algorithm uses 1520 *a priori* profiles that vary with latitude, total O_3, and month. Shown are some examples of these *a priori* ozone profiles. The altitude at which the ozone partial pressure peaks drops with latitude and total O_3.

Lambertian reflectors. However, they recognized that for the retrieval of total O_3 one uses contrast in (top-of-the-atmosphere) reflectances (TOAR) with wavelength caused by O_3, rather than the absolute reflectance; therefore, the key issue is not what the LER represents but how does it vary with wavelength. They considered this issue so important that the original TOMS instrument (launched in Oct. 1978 on the Nimbus-7 satellite) was built with 4 wavelengths (331.2, 340, 360, and 380 nm) from which LERs could be derived and compared. By contrast, the instrument had only two O_3 sensitive wavelengths (312.5 and 317.5 nm)! In more recent versions of the TOMS algorithm, as well as in most versions of the DOAS algorithms, the LER concept has been modified to handle mixed cloudy scenes better by treating such scenes as mixtures of two Lambert surfaces at two different altitudes; hence called the mixed LER model (MLER). Using Mie scattering theory, Ahmad et al. [2004] show that the MLER model also handles thin clouds better than does the LER model (except over very bright snow-covered scenes).

In the basic TOMS version 8 algorithm [Wellemeyer et al., 2004] a weakly ozone-absorbing wavelength (mostly 331.2 nm, but 360 nm at very large SZAs) is used to derive either the LER (when the atmosphere is deemed clear, fully cloudy, or when the ground is covered with snow or ice) or the cloud fraction (f_c). (One starts with a nominal total O_3 amount to correct for the weak ozone absorption at 331.2 and 360 nm, which is then refined as a better estimate of total ozone becomes available.) Then, assuming that LER/f_c is wavelength independent, one can calculate the TOAR (ρ) at any wavelength using an accurate radiative transfer program [Dave, 1964], an *a priori* ozone profile, and a temperature profile. Total ozone can then be obtained by solving the "Option 1" formula in Figure 7 iteratively, starting with a first guess. However, given limited number of O_3 sensitive wavelengths, the TOMS algorithm solves this equation without the polynomial term on the right side of equation. Thus, total O_3 is derived from the last term only, using just a single wavelength (mostly 317.5 nm, 331.2 nm at large SZAs). Based on the previous description of the DS technique, it can be seen that this method has the same advantage over the DOAS methods as does the Dobson–Brewer technique; both use the much larger difference in total absorption (between 317.5 and 331.2, or between 331.2 and 360 nm at large SZAs) rather than the residual absorption used in the DOAS methods; this simplifies the instrument design—requiring as few as three wavelengths to retrieve total O_3. However, without the polynomial term, the TOMS algorithm is more sensitive to wavelength dependence of LER and instrument radiometric drifts. We discuss these issues below.

Wavelength Dependence of LER: Note that the basic TOMS algorithm does not do an explicit correction for aerosols. The rationale for this was provided by Dave [1978]. By explicit Mie scattering calculation, Dave showed that many common types of (non-UV-absorbing) aerosols simply increase the LER of the surface without modifying its spectral behavior, hence they produce no error in the derivation of O_3. However, in the same paper Dave showed that aerosols that have strong UV absorption would cause the LER to decrease with wavelength, thus such aerosols would introduce error in deriving total O_3. Unfortunately, because of interference from subpixel clouds this effect was not discovered in the TOMS data until 1992, 14 years after launch of the first TOMS instrument, when the LER model was replaced with the MLER model, which removed the cloud interference. It is now well established that smoke, desert dust, and volcanic ash strongly absorb UV radiation, and they do produce the effect Dave had predicted. Indeed this observation has led to a unique new way of detecting such aerosols from UV satellite data. The so-called aerosol index, which is derived from the difference in LER between two TOMS wavelengths, is now popularly used for tracking such aerosols over the globe [Hsu et al., 1996; Herman et al., 1997]. (Figure 10 shows an example of how one can track smoke using the TOMS AI) Since these aerosols do produce error in the TOMS algorithm, a correction to the basic TOMS total O_3 is applied using the Aerosol Index (AI) as a predictor. (See Torres & Bhartia, 1999 for an earlier version of this correction procedure. The functional form of the correction is somewhat more complex in the TOMS version 8 algorithm.)

It is worth mentioning that besides UV-absorbing aerosols highly non-Lambertian surfaces, such as the water surface viewed in the geometrical reflection direction (glint), can also cause LER to vary with wavelength. This happens because the LER of such surfaces is very different for direct and diffuse light; from satellite one measures the mixture of the two LERs weighted by the ratio of direct to diffuse radiation. Since this ratio varies with wavelength, so does the

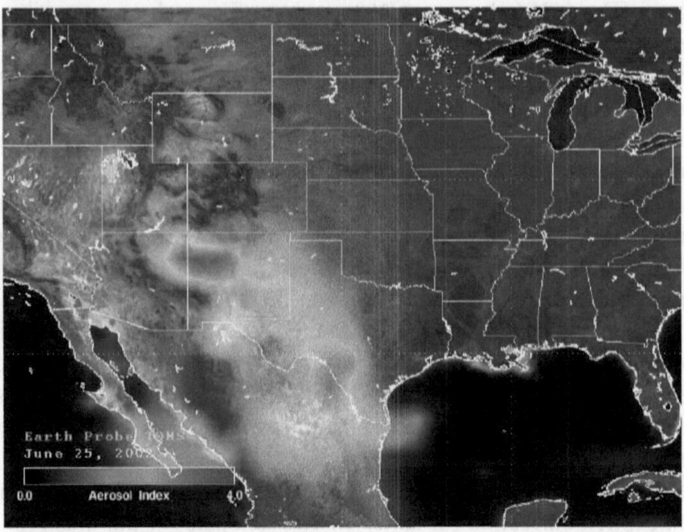

FIGURE 10. This false-color map of TOMS aerosol index (AI) shows smoke from fires in the western US. AI is produced from the difference in LER between two TOMS wavelengths. The color of the land in the background is a visible RGB image from another sensor. Note the absence of clouds in the picture; the TOMS AI algorithm is not sensitive to clouds.

LER. When the surface of the water is viewed within ~20° of the glint direction, the LER of water surfaces decreases sharply with wavelength, as the direct to diffuse ratio decreases. The magnitude of this effect varies strongly on the surface wind speed, aerosols, and clouds. The AI also allows one to measure the strength of the glint and to correct for its effect on TOMS O_3.

Radiometric Drift: The radiometric drift (change in instrument sensitivity to incoming radiation) of the instrument can affect the TOMS algorithm in several ways. First, it can cause the derived LER to drift, second it can cause the TOARs (ρ) at the O_3 wavelength (317.5 nm) to drift, and third it can cause the AI to drift. All these effects will cause the estimated O_3 to drift. Over the years robust techniques have been developed to monitor these drifts. For example, the drift in LER can be monitored by looking at the LER of Greenland and/or Antarctica, both of which are quite stable (though their LERs varies with SZA). The drift in AI can be monitored by ensuring that the AI of bright clouds remains zero with time. Monitoring the drift of the TOAR at 317.5 nm is more difficult. Though several internal techniques have been developed over the years to check the drift in the 317.5 nm TOAR [Herman et al., 1991], ultimately, comparisons with Dobson–Brewer instruments are necessary to maintain data quality over the long-term. This is one area in which the DOAS algorithms may prove superior to the TOMS algorithm.

Key Results from TOMS

Figure 11 shows that the first TOMS instrument lasted about 13.5 years [McPeters et al., 1996]. It was followed by two other TOMS instruments. EP/TOMS is currently operating, though it suffers from significant calibration problems that have degraded the data quality (Another TOMS instrument, not shown in Figure 10, was launched on the Japanese ADEOS satellite in 1996, around the same time that EP/TOMS was launched. Since the satellite stopped working after about 9 months, data from this instrument have not been studied extensively.)

Figures 12 and 13 show what is generally considered to be the two most important results derived from the TOMS data. Figure 12 shows the long-term changes in the area weighted total ozone between 60S–60N. By combining the data from the both hemispheres, one removes most of the out-of-phase seasonal variation in total O_3; some variation that remains is caused by the fact that the amplitude of the seasonal cycle is stronger in the NH than in the SH. Long-term decrease in global total O_3 is clearly scene in this data. Although some of this decrease could be due to changes in the meridional transport from tropics to higher latitudes, the bulk of this decrease is believed to be due to increase in the halogen loading of the stratosphere, which causes ozone depletion. The dotted line is a prediction using a two-dimensional stratospheric chemistry model developed at NASA Goddard Space Flight Center. The inter-annual oscillations seen in the model are caused by the effects of solar cycle and changes in aerosol

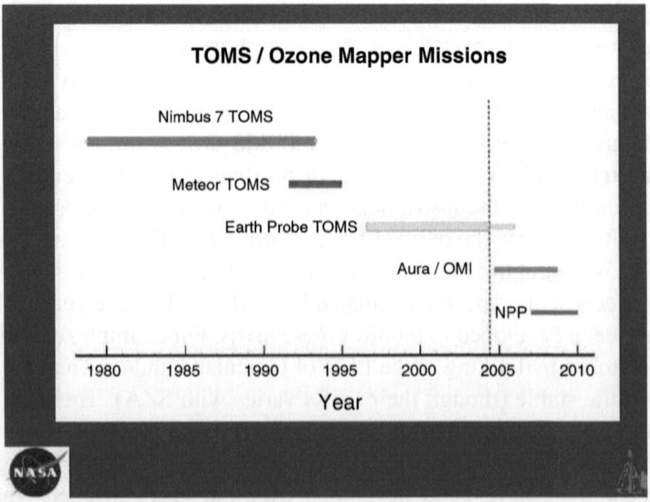

FIGURE 11. The TOMS O_3 record started in Oct. 1978 and now spans a quarter-century. More advanced instruments, such as OMI, launched recently on July 15, 2004, on the EOS Aura satellite, will continue this record.

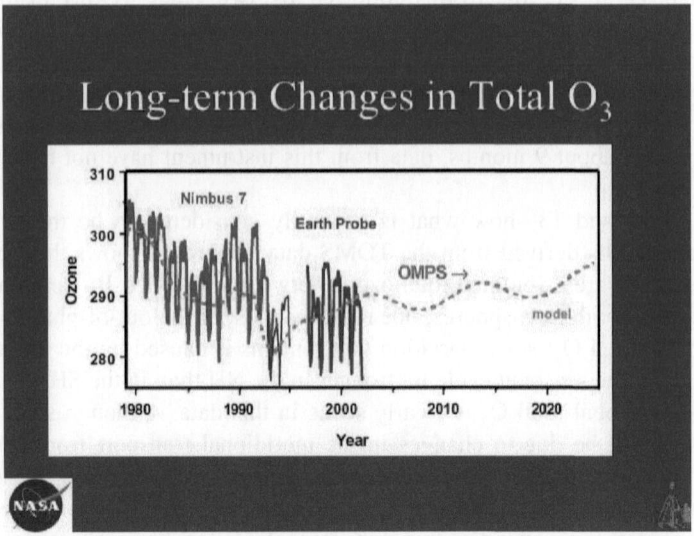

FIGURE 12. Long-term change in global (60S–60N) total O_3 monitored by TOMS. The dotted line is a model prediction.

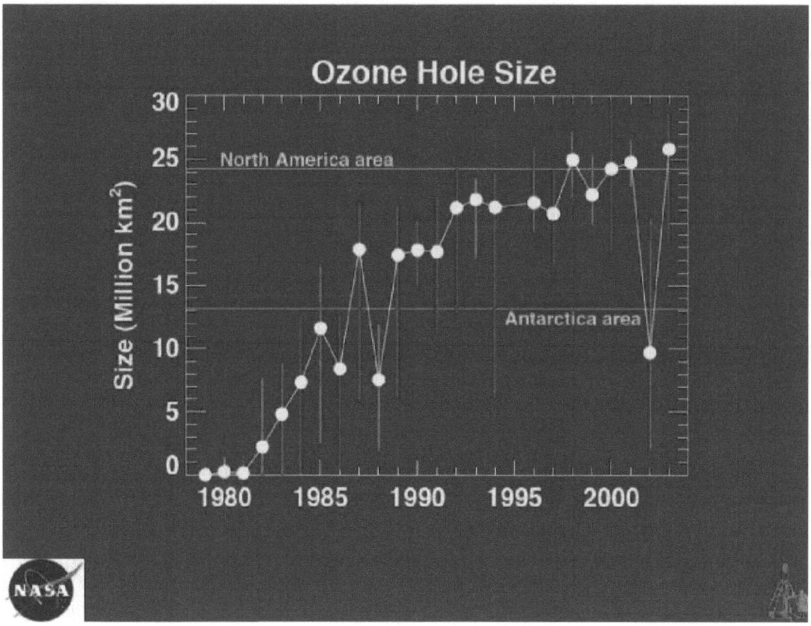

FIGURE 13. Evolution of the area of the Antarctica O_3 hole as seen by TOMS.

loading after the eruption of two major volcanoes in the last quarter century: El Chichon in Mexico and Mt. Pinatubo in the Philippines. Figure 13 shows the evolution of the area of the Antarctica ozone hole as seen by TOMS. Though the area varies substantially from year-to-year due to variations in the size and strength of the polar vortex, the long-term increase in area is due to ozone depletion caused by man-made chemicals.

Summary

The first paper describing a method for estimating total column O_3 from satellite backscatter measurements was published 39 years ago. With advances in computer power, better understanding of atmospheric radiative transfer, and improved instrument design, it is now possible to derive total O_3 from satellites with better than 2% accuracy, and ~1% precision over conditions ranging from clear to fully cloudy, up to 85° solar zenith angle, and in presence of massive aerosol loading. There are three basic types of algorithms currently in use. They are briefly described below.

Discrete Wavelength Algorithm: This algorithm, derived from Dave and Mateer [1967], has been used to process a quarter-century of TOMS data. The basic algorithm uses only two wavelengths, but 2–3 additional wavelengths are necessary to correct for errors due to aerosols, sea glint, and volcanic SO_2. With proper selection of wavelengths the effect of atmospheric temperature can be minimized. This algorithm has the advantage that it can be applied to simple instruments that have few spectral channels with relatively coarse spectral resolution (1–2 nm). However, the algorithm is sensitive to instrument drift, which must be monitored by using external reference instruments.

Spectral-Fitting Algorithms: These algorithms offer reduced sensitivity to instrument drift but require instruments with narrow bandpass (<0.5 nm), and multiple spectral samples per bandpass, to minimize systematic errors. There are two types of spectral-fitting algorithms. Those that use narrow spectral windows (< 10 nm) are based on the DOAS technique. More advanced algorithms using a broader window (15–20 nm) are being developed to achieve higher accuracy/precision.

Maximum-Likelihood Estimation Algorithm: These algorithms derive ozone profile using a broad set of wavelengths (280–320 nm) from which total O_3 is obtained by vertical integration. They offer reduced sensitivity to *a priori* O_3 profiles, which becomes a significant error source for the previous two algorithms when solar zenith angle exceeds 80°. However, these algorithms require instruments that can measure weak UV signals down to 280 nm with minimum of scattered/stray light contamination from longer wavelengths.

References

Ahmad, Z., P.K. Bhartia, N. Krotkov (2004), Spectral properties of backscattered UV radiation in cloudy atmospheres, *J. Geophys. Res.*, **109**, D01201, doi:10.1029/2003JD003395.

Bates, D.R. (1984), Rayleigh scattering by air, *Planet. Sp. Sci.*, **32**, 785–790.

Bhartia, P.K., C.G.Wellemeyer, S.L. Taylor, N. Nath and A. Gopalan (2004), Solar Backscatter Ultraviolet (SBUV) Version 8 Profile Algorithm, Proceeding Quadrennial Ozone Symposium, 1–8 June 2004, Kos, Greece.

Dave, J.V. (1964), Meaning of successive iteration of the auxiliary equation of radiative transfer, *Astrophys. J.*, **140**, 1292–1303.

Dave, J.V. and C.L. Mateer (1967), A preliminary study on the possibility of estimating total atmospheric ozone from satellite measurements, *J. Atmos. Sci.*, **24**, 414–427.

Dave, J.V. (1978), Effect of aerosols on the estimation of total ozone in an atmospheric column from the measurement of its ultraviolet radiance, *J. Atmos. Sci.*, **35**, 899–911.

Dobson, G.M.B. (1931), A photo-electric spectrometer for measuring the amount of atmospheric ozone, Proc. Phys. Soc., 324–339.

M. Coldewey-Egbers, M. Weber, L. N. Lamsal, R. de Beek, M. Buchwitz, J., P. Burrows (2004), Total ozone retrieval from GOME UV spectral data using the weighting function DOAS approach, Atmos. Chem. Phys. Discuss., **4**, 4915–4944.

Herman, J.R., R. Hudson, R. McPeters, R. Stolarski, Z. Ahmad, X.Y. Gu, S. Taylor, and C. Wellemeyer (1991), A new self-calibration method applied to TOMS and SBUV backscattered ultraviolet data to determine long-term global ozone changes, *J. Geophys. Res.*, **96**, pp. 7531–7545.

Herman, J.R., P.K. Bhartia, O. Torres, C. Hsu, C. Seftor and E. Celarier (1997), Global Distribution of UV-absorbing aerosols from Nimbus7/TOMS data, *J. Geophys. Res.*, **102**, 16911–16922.

Hsu, N.C., J.R. Herman, P.K. Bhartia, C.J. Seftor, O. Torres, A.M. Thompson, J.F. Gleason, T.F. Eck, and B.N. Holben (1996), Detection of biomass burning smoke from TOMS measurements, *Geophys. Res. Lett.*, **23**, 745–748.

Klenk, K.F., P.K. Bhartia, A.J. Fleig, V.G. Kaveeshwar, R.D. McPeters, and P.M. Smith (1982), Total ozone determination from the Backscattered Ultraviolet (BUV) experiment, *J. Appl. Met.*, **21**, 1672–1684.

McPeters, R.D., et al. (1996), Nimbus7 Total Ozone Mapping Spectrometer (TOMS) Data Product's User's Guide, *NASA Reference Publication 1384*, National Aeronautics and Space Administration, Washington, DC.

Rodgers, C.D. (2000), Inverse methods for Atmospheric Sounding: Theory and Practice, World Scientific Publishing Co. Ltd.

Torres, O. and P.K. Bhartia (1999), Impact of tropospheric aerosol absorption on ozone retrieval from backscattered ultraviolet measurements, *J. Geophys. Res.*, **104**, 21569–21577.

Valks, P.J.M., J.F., de Haan, J.P. Veefkind, and R.F., van Oss, TOGOMI (2004), An improved total ozone retrieval algorithm for GOME, Proceeding Quadrennial Ozone Symposium, 1–8 June 2004, Kos, Greece.

Wellemeyer, C.G., S.L. Taylor, C.J. Seftor, R.D. McPeters and P.K. Bhartia (1997), A correction for total ozone mapping spectrometer profile shape errors at high latitude, *J. Geophys. Res.*, **102**, 9029–9038.

Wellemeyer, C.G., P.K. Bhartia, S.L.Taylor, W. Qin, and C. Ahn (2004), Version 8 total ozone mapping spectrometer (TOMS) algorithm, Proceeding Quadrennial Ozone Symposium, 1–8 June 2004, Kos, Greece.

Yang, K., P.K. Bhartia, C.G. Wellemeyer, W. Qin, R.J.D. Spurr, J.P. Veefkind, J.F. De Haan (2004), Application of spectral fitting method to GOME and comparison with OMI DOAS and TOMS-V8 total ozone, Proceeding Quadrennial Ozone Symposium, 1–8 June 2004, Kos, Greece.

Chapter 4
The EOS Aura Mission*

MARK R. SCHOEBERL

The Earth Observing System Missions and Aura

The Earth Observing System (EOS) Aura satellite mission launched July 15, 2004. The Aura mission will make significantly improved measurements atmospheric constituent. Aura is designed to attack three science questions: (1) Is the ozone layer recovering as expected? (2) What are the sources and processes that control tropospheric pollutants? (3) What is the quantitative impact of constituents on climate change? Aura will answer these questions by globally measuring a comprehensive set of trace gases and aerosols (Table 1) at high vertical and horizontal resolution. Figure 1 shows the Aura spacecraft and its four instruments.

The EOS Program consists of three core satellites, Terra (http://eos-am.gsfc.nasa.gov/), Aqua (http://eos-pm.gsfc.nasa.gov/), and Aura (http://eos-aura.gsfc.nasa.gov/) as well as several smaller satellites. Aura (Latin for "breeze") was launched into an ascending node 705 km sun-synchronous polar orbit with a 98 ° inclination with an equator-crossing time of 13:45±15 minutes. The design life is five years with an operational goal of six years. Aura will fly in the same orbit track about 15 minutes behind Aqua, the Cloud-Aerosol Lidar and Infrared Pathfinder Satellite Observation (CALIPSO, http://www.calipso.larc.nasa.gov/) and Cloudsat launched on 28 April 2006 [Stephens et al., 2002, http://cloudsat.atmos.colostate.edu/]. This group of satellites, including the CNES PARASOL satellite (http://smsc.cnes.fr/PARASOL/GP_mission.htm) and the ESSP Orbiting Carbon Observatory (OCO, http://oco.jpl.nasa. gov/), is referred to as the "A-Train." The measurements from Aura will be within 30 minutes of these other platforms. The near simultaneous measurements by the A-Train satellites will improve estimates of aerosol and cloud interaction and their role in precipitation and climate forcing.

---

* This is a slightly modified version a paper that appeared in EOS [Schoeberl et al., 2004].

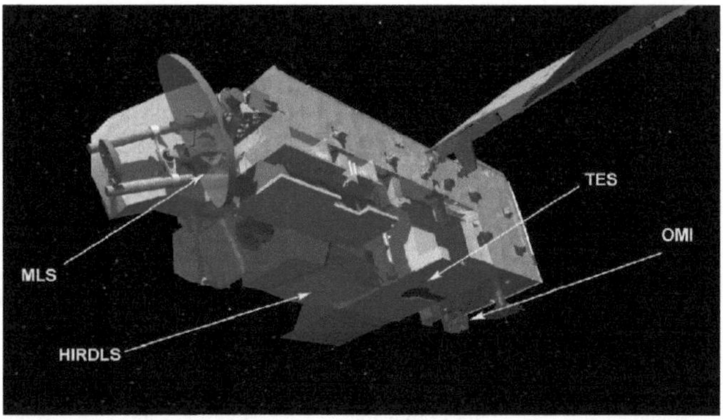

FIGURE 1. Computer model of the Aura spacecraft showing the location of HIRDLS, MLS, OMI and TES. See Table 1. (Graphic by Jesse Allen, NASA Earth Observatory)

Science Objectives of the Aura Mission

When combined with field campaign data, other satellite measurements, and ground-based observations, Aura measurements will provide unprecedented insights into atmospheric chemical and dynamical processes.

Is the Ozone Layer Recovering as Expected?

Total Ozone Mapping Spectrometer (TOMS). Observations from 1978 show strong secular decrease in column ozone at extra-tropical latitudes. Although the Antarctic ozone hole area growth has slowed, significant late winter ozone depletions have now occurred in the Arctic [WMO, 2002]. Upper Atmosphere Research Satellite (UARS) data show a flattening in the stratospheric chlorine concentrations [Anderson et al., 2000]. A decrease in chlorine should lead to recovery of the ozone layer, but this recovery may be altered because of increasing greenhouse gas cooling [e.g., Shindel and Grewe, 2002]. As a result of the uncertainty in stratospheric trace gas trends, temperatures and dynamical feedback processes, current models used to assess the ozone layer do not agree on the timing of the ozone layer recovery [WMO, 2002].

The stratospheric measurements made by Aura will permit a complete assessment of the chemical processes controlling ozone. First, high vertical resolution ozone profiling by MLS and HIRDLS will provide the best information ever on ozone change. Second, five of the major radicals that participate in ozone destruction (ClO, OH, HO_2, BrO, and NO_2) will be measured by either HIRDLS or MLS. Third, MLS and HIRDLS also measure the important reservoir gases, HCl, $ClONO_2$, and HNO_3. Fourth, the Aura instrument payload will make measurements of the long-lived source gases including N_2O, H_2O, CH_4, CFC's. Finally,

TABLE 1. Aura Instruments, Principle Investigators and Measurements (Most measurements will have a precision of 10% or better.)

Acronym	Name	Instrument PI	Constituent	Instrument Description
HIRDLS	High Resolution Dynamics Limb Sounder	John Gille, National Center for Atmospheric Research & U. of Colorado; John Barnett, Oxford University	Profiles of T, O_3, H_2O, CH_4, N_2O, NO_2, HNO_3, N_2O_5, CF_3Cl, CF_2 Cl_2, $ClONO_2$, Aerosols	Limb IR filter radiometer from 6.2μ to 17.76μ 1.2 km vertical resolution up to 80 km.
MLS	Microwave Limb Sounder	Joe Waters, Jet Propulsion Laboratory	Profiles of T, H_2O, O_3, ClO, BrO, HCl, OH, HO_2, HNO_3, HCN, N_2O, CO, Cloud ice.	Microwave limb sounder 118 GHz to 2.5 THz 1.5–3 km vertical resolution
OMI	Ozone Monitoring Instrument	Pieternel Levelt, KNMI, Netherlands	Column O_3, SO_2, aerosols, NO_2, BrO, OClO. HCHO, UV-B, cloud top pressure, O_3 profiles.	Hyperspectral nadir imager, 114 ° FOV, 270-500 nm, 13×24 km footprint for ozone and aerosols
TES	Tropospheric Emission Spectrometer	Reinhard Beer, Michael Gunson, Jet Propulsion Laboratory	Profiles of T, O_3, NO_2, CO, HNO_3, CH_4, H_2O.	Limb (to 34 km) and nadir IR Fourier transform spectrometer 3.2–15.4μ Nadir footprint 5.3×8.5 km, limb 2.3 km

OMI will continue the TOMS/SBUV global column and profile ozone trend measurements.

What Are the Sources and Processes That Control Tropospheric Pollutants?

Tropospheric ozone production occurs when CO, volatile organic compounds (VOCs), and nitrogen oxides are exposed to sunlight. These ozone precursors are directly linked to urban sources, and the atmosphere can transport both ozone and its precursors over large distances. The Aura mission is designed to produce the first global assessment of tropospheric ozone and its precursors, as well as assess the stratospheric contribution to the tropospheric ozone budget. The measurements from TES as well as column measurements from OMI combined with stratospheric measurements from HIRDLS and from MLS will provide new information on the pollution sources and transport.

What Is the Quantitative Impact of Aerosols and Upper Tropospheric Water Vapor and Ozone in Climate Change?

One of the sources of uncertainty in climate change comes from our poor understanding of the changing concentration of upper troposphere and lower stratospheric (UT/LS) constituents [Houghton et al., 1995]. On decadal time scales, climate signals from greenhouse gas changes and changes in reactive constituents are intertwined. Aura's MLS and HIRDLS instruments have been specifically designed to study the UT/LS in unprecedented detail. MLS can make constituent measurements through thin clouds while HIRDLS can scan vertically across the limb over a wide horizontal range through cloud gaps. The issues of climate radiative forcing cannot be solved by Aura alone. Data from Aqua, Cloudsat, and CALIPSO will also be needed to address climate issues fully.

Aura Instrument Descriptions

Figure 2 shows the instrument fields of view. MLS will make limb sounds observing forward. OMI and TES will make nadir soundings. HIRDLS and TES will make limb soundings observing backward. The advantage of this instrument configuration is that each of the instruments can observe the same air mass within ~13 minutes.

FIGURE 2. Aura instrument instantaneous fields of view, looking toward the back of the spacecraft. MLS limb measurements, green, OMI nadir measurements, blue swath. TES limb and nadir measurements, red; HIRDLS measurements yellow. Limb sounders scan vertically. (Graphic by Jesse Allen, NASA Earth Observatory.)

HIRDLS

HIRDLS is an infrared (6–1 μm) limb scanning, 26-channel filter radiometer. HIRDLS makes temperature and trace gases observations from the UT/LS to the mesosphere (Table 1) [Gille et al., 2003]. HIRDLS was jointly built by the United Kingdom and the United States. HIRDLS can also detect clouds and thus determine the altitude of polar stratospheric clouds and tropospheric cloud tops. HIRDLS has higher horizontal resolution than any previous limb sounder through the use of a programmable azimuth scan in conjunction with a rapid elevation scan. Special observing modes can be used to observe geophysical events like volcanic clouds.

MLS

MLS is a 118 GHz–2.5 THz, limb scanning microwave emission spectrometer that measures temperature and constituents from the UT/LS to the mesosphere (Table 1) [Waters et al., 1999]. MLS also has capability of measuring upper tropospheric water vapor in the presence of cirrus. Aura MLS is based on UARS MLS but uses more advanced technology to make additional constituent measurements. These new constituents play an important role in stratospheric chemistry (i.e., HCl, N_2O, OH, HO_2 and BrO). The UARS MLS was able to measure upper tropospheric water vapor [Read et al., 2001], which is essential for understanding climate variability. Aura MLS will improve this measurement.

OMI

The OMI instrument is a contribution of the Netherlands' Agency for Aerospace Programs (NIVR) in collaboration with the Finnish Meteorological Institute (FMI) [Levelt et al., 2000]. OMI will continue the TOMS record for total ozone and other atmospheric parameters related to ozone chemistry and aerosols. OMI is a 0.24–0.50-μm, visible-UV 740 band, cross track hyperspectral imager. OMI will provide global coverage in one day at 13 × 24 km spatial resolution. A combination of backscatter ultraviolet retrieval algorithms and forward modeling will be used to generate the various OMI data products [Ahmad et al., 2003].

TES

TES is a 3.2–15.4 μm, high-resolution infrared-imaging Fourier Transform spectrometer. TES has a spectral resolution of 0.025 cm^{-1} measuring most of the radiatively active molecular species in the Earth's atmosphere [Beer et al., 2001]. TES can make both limb and nadir observations. TES can target within 45 ° of the local vertical, or produce regional transects over clear regions up to 885 km in length. TES will provide global maps of trace gases listed in Table 1. Because TES measures the entire IR spectrum, the potential exists to retrieve a large number of other gases (e.g., ammonia); although the retrieval of these gases will be done in a research mode.

Aura Instrument Synergy

Because Aura instruments observe the same air mass within ~13 minutes, it will also be possible to better interpret photochemical processes involving constituents measured simultaneously as has been done with UARS data (e.g., Douglass et al., 1995). In addition, HIRDLS and MLS limb sounding will provide ozone profiles that can be combined with the OMI column ozone observations. It will then be possible to separate the stratospheric component of the column ozone and thus estimate the tropospheric ozone column. The tropospheric column can be compared with the TES direct measurement of tropospheric ozone.

Summary

The five-year EOS Aura mission will provide a significantly improved tropospheric and stratospheric constituent measurements. Although there are only four instruments on Aura, the breadth of the their capability combined with the other A-Train measurements will provide a powerful tools to attack future questions about changing atmospheric composition and its impact on climate. The instruments OMI, MLS, and TES are on and performing nominally. Observations of the 2004 Antarctic ozone hole have already been made by MLS and OMI. HIRDLS is experiencing radiative anomalies due to a piece of kapton blocking the aperture; its future performance is uncertain.

References

Anderson, J.J., et al., 2000: Halogen occulation experiment confirmation of stratospheric chlorine decreases in accordance with the Montreal protocol, *J. Geophys. Res.*, **105**, 4483–4490.

Ahmad, S.P., et al. Atmospheric products from the ozone monitoring instrument (OMI), Proceedings of SPIE July 2003, San Diego, U.S.A., 5151–66, 2003.

Beer, R., T.A. Glavich and D.M. Rider, Tropospheric emission spectrometer for the Earth Observing System's Aura satellite. Applied Optics, **40**, 2356–2367, 2001.

Douglass, A.R. et al., Interhemispheric Differences in Springtime Production of Vortex Chlorine Reservoirs, *J. Geophys. Res.*, **100**, 13967–13978, 1995.

Gille, J., et al., A. Lambert and W. Mankin, The High Resolution Dynamics Limb Sounder (HIRDLS) Experiment on Aura, Proc. of SPIE, **5152**, 162–171, 2003.

Houghton, J.T., et al. (eds.), *Climate Change 1994, The IPCC Scientific Assessment*, Cambridge University Press, 339 pp., 1995.

Levelt, P.F., et al. *Science Objectives of EOS-AURA's Ozone Monitoring Instrument (OMI)*, Proceedings of the Quadrennial Ozone Symposium, Sapporo July 2000, pp. 127–128, 2000.

Read, W.G., et al., UARS Microwave Limb Sounder Upper Tropospheric Humidity Measurements: Method and Validation. *J. Geophys. Res.*, **106**, 32,207–32,258, 2001.

Schoeberl et al. Earth Observing Systems Benefit Atmospheric Research, EOS, **85**, 177–178, 2004.

Shindell, D.T. and V. Grewe, Separating the influence of halogen and climate change on ozone recovery in the upper stratosphere, J. Geophys. Res., **107**, doi:10.1029/2001 JD000420, 2002.

Stephens, G.L., et al., The CloudSat mission and the A-train—A new dimension of space-based observations of clouds and precipitation, *Bull. Amer. Met. Soc.*, **83**, 1771–1790, 2002.

Waters, J.W., et al., The UARS and EOS Microwave Limb Sounder (MLS) Experiments. *J. Atmos. Sci.*, **56**, 194–218, 1999.

World Meteorological Organization, Scientific Assessment of Ozone Depletion: 2002, Report No. 47, 2002.

Chapter 5
MIPAS Experiment Aboard ENVISAT

HERBERT FISCHER

Introduction

At March 1, 2002, the European Space Agency (ESA) launched the ENVISAT satellite in a polar, sun-synchronous orbit. The large satellite with a mass of more than eight tons is orbiting in a height of about 800 km with a descending node of 10:00 h. From the ten experiments on board three are used to explore the composition of the atmosphere. One of them is the Michelson Interferometer for Passive Atmospheric Sounding (MIPAS).

The Institute for Meteorology and Climate Research (IMK) has developed in the past several MIPAS instruments that have been used in ground stations, aboard aircrafts and balloon gondolas [Fischer, 1993]. Finally, the IMK has proposed a corresponding satellite experiment to ESA who has included this MIPAS experiment in the core payload of ENVISAT (see Figure 1).

MIPAS/ENVISAT Experiment

MIPAS is a high-resolution Fourier Transform Spectrometer that measures the atmospheric limb emission simultaneously in four spectral bands in the middle infrared [H. Fischer et al., 2000]. The whole spectral interval covers the range from 4.15 μm to 14.6 μm. The unapodized spectral resolution is 0.035 cm^{-1}, which allows the detection of many single spectral emission lines. MIPAS is continuously performing measurements during day and night, providing each day a full coverage of the globe. During an orbit the instrument performs 75 limb scans and in addition calibration measurements when operating in the nominal mode. The instrument is very flexible, i.e., it can also be used for performing a considerable number of other observation modes with various scientific objectives.

By analysing a limb scan sequence vertical profiles of temperature and many trace gases can be derived. Furthermore, these data are appropriate to determine information about aerosols as well as particles of Polar Stratospheric and cirrus clouds.

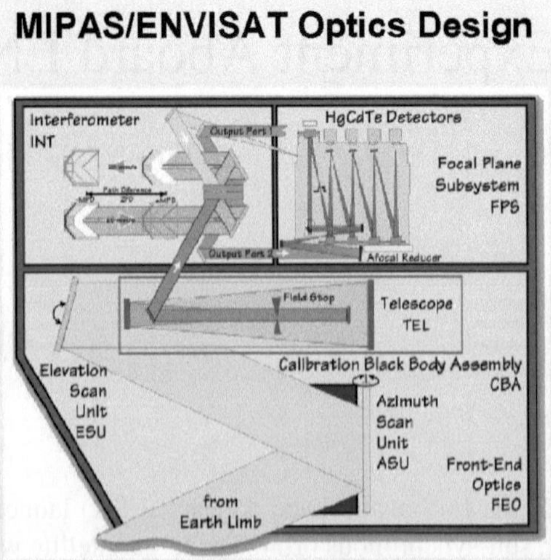

FIGURE 1. Optical design of the MIPAS/ENVISAT experiment.

Combining the profiles along an orbit, the distribution of the atmospheric parameters can be presented as a function of altitude and latitude.

The MIPAS instrument is a dual-slide, dual-port interferometer that is using all the incoming radiance for measurement and is in various instrumental parts redundant. Figure 1 shows the optical design of MIPAS-ENVISAT including the azimuth and elevation scan unit, the telescope, the interferometer, and the focal plane subsystem with the detectors. Table 1 summarizes the main specifications of the MIPAS experiment.

TABLE 1. Summary of MIPAS Specifications

ENVISAT launch: 1st March 2002	
Michelson interferometer	
Spectral range: 4.1–14.6 μm	
Spectral resolution: 0.035 cm^{-1}	
Altitude range:	
Nominal mode: 6–68 km	
several other observation modes	
Global coverage: measurements from pole to pole	
Measurement schedule: continuous	
Measurement parameters:	
-p, T, H_2O, O_3, CH_4, N_2O, HNO_3, NO_2	operational processing
-NO, N_2O_5, HNO_4, $ClONO_2$, ClO, CO, CFCs, NH_3, C_2H_6	scientific processing
HDO, O_3 isotopomers and others (more than 25 parameters)	

The enormous amount of information in one MIPAS spectrum is shown in Figure 2. This emission spectrum of the atmosphere associated to a tangent altitude of 26 km (definition see below in Fig. 4) impresses by the large number of spectral lines which stem from many atmospheric trace gases as indicated in the figure. Most prominent are the absorption bands of CO_2 (15 μm), O_3 (9.6 μm), CH_4 (7.7 μm), and H_2O (6.3 μm).

MIPAS-Experiment Aboard ENVISAT

Data Processing

An overview of the measurement and data processing is given in Figure 3. The interferometer transforms the atmospheric radiance in an interferogram, which has to be converted back to a spectrum by applying a Fourier transformation. In the following this spectrum has to be calibrated by using blackbody and deep space measurements. A further step in data processing is the inversion of spectral radiances in trace gas concentrations. A trace gas profile can be derived from a sequence of limb measurements.

The limb sounding geometry is shown in Figure 4. The ray path is mainly defined by the tangent altitude, which corresponds to the minimum distance between ray path and Earth surface. The information about the vertical structure of the atmosphere is attained by scanning across the atmosphere. The Radiative

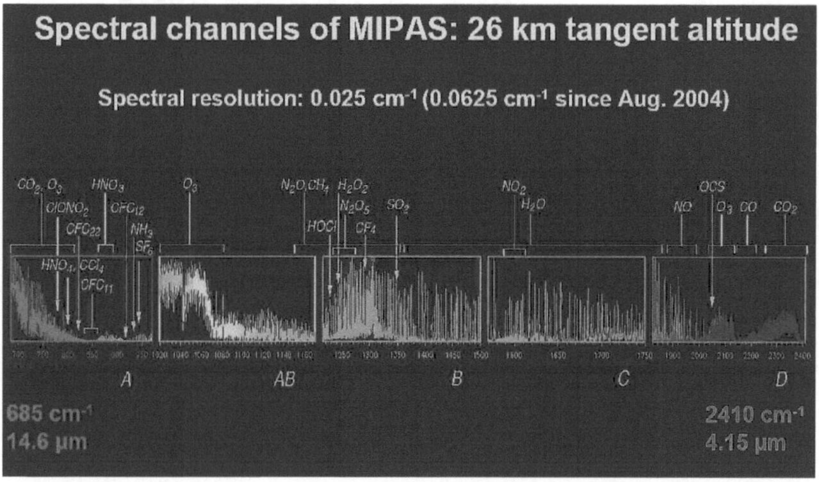

FIGURE 2. MIPAS spectrum for a tangent height of 26 km showing the enormous amount of information contained.

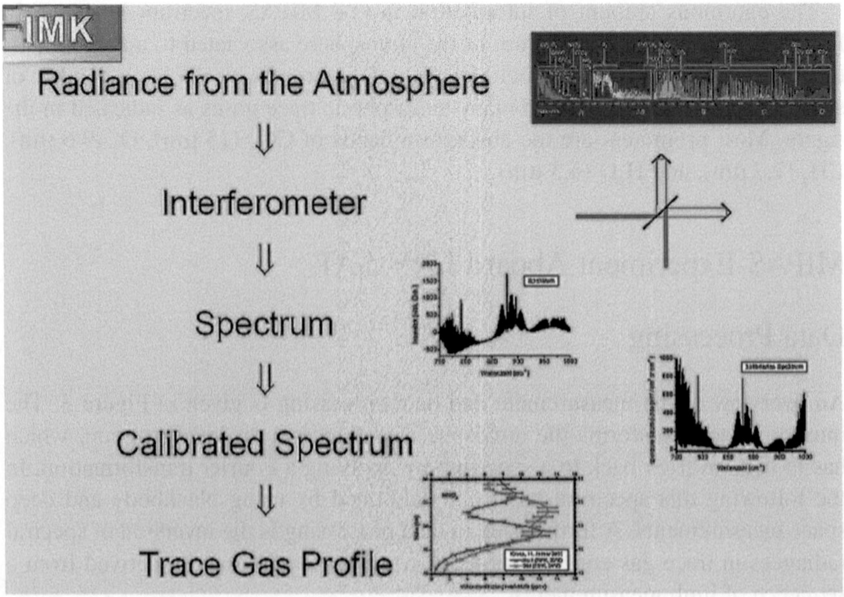

FIGURE 3. Overview of the measurement and data processing.

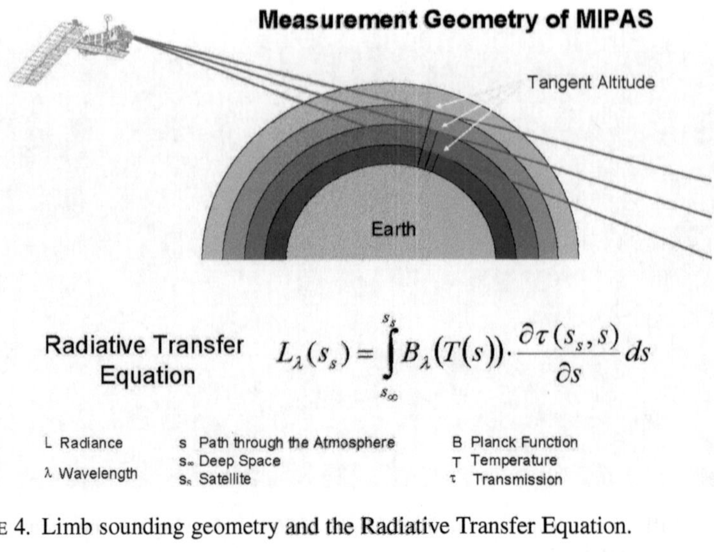

$$L_\lambda(s_s) = \int_{s_\infty}^{s_s} B_\lambda(T(s)) \cdot \frac{\partial \tau(s_s, s)}{\partial s} ds$$

FIGURE 4. Limb sounding geometry and the Radiative Transfer Equation.

Transfer Equation (RTE) describes the connection between the measured radiances and the distribution of the atmospheric parameters. A corresponding computer code is called KOPRA (Stiller et al., 2000), which has been developed at the IMK. It is based on the up-to-date knowledge about the radiative processes in the atmosphere including the ray tracing for an elliptical earth, the consideration of horizontal inhomogeneities, and non-LTE (local thermodynamic equilibrium) effects (Table 2).

In order to determine vertical profiles of atmospheric parameters from the radiance measurements, the RTE has to be inverted (i.e., the integral equation has to be solved). Methods for this so-called retrieval are described by Rodgers [2000]. We are using the optimal estimation method with a Tikhonov smoothing operator.

A sequence of limb emission measurements (between 68.4 km and 6.4 km tangent altitude) is shown in Figure 5. First of all, a gap of information can be recognized in the spectral range between 970 and 1020 cm^{-1}. At high altitudes the emitted radiance is relatively weak; only the CO_2-band below 760 cm^{-1} and the O_3-band around 1050 cm^{-1} are clearly visible. With decreasing tangent altitude the emitted radiance becomes stronger and additional absorption bands show up, e.g., the CO_2 laser band around 960 cm^{-1}, the HNO_3-bands around 875 cm^{-1}, and the CFC bands around 850 cm^{-1} and 920 cm^{-1}. The HNO_3 bands grow rapidly in the lower stratosphere due to the shape of the vertical profile of the HNO_3 mixing ratio. In the upper troposphere the spectrum exhibits some single strong lines in the range of the atmospheric window between 800 and 1000 cm^{-1}. They are caused by weak water vapor lines becoming visible in the upper troposphere due to the quickly increasing water vapor mixing ratio.

TABLE 2. Main Characteristics of the Radiative Transfer Code KOPRA

Main Characteristics of KOPRA
• Currently ~70 different trace species, many of them relevant in the UTLS only
• Non-LTE radiative transfer modelling
• Vertical variation of isotopomeric abundances
• Accurate modelling of spectroscopic features (CO_2, line mixing, self/air broadening, pressureshift, line shapes, continua, p/T-dependent cross-sections of heavy molecules, ...)
• Particle-caused absorption, emission <u>and</u> single scattering
• Pseudo-analytic derivatives
• Generic non-LTE state population model coupled into KOPRA
• Mie model for calculation of absorption and scattering coefficients and phase function coupled into KOPRA
• Raytracing for elliptical earth
• Consideration of horizontal inhomogeneities (P, T, vmr, ...) by 3-D fields
• Derivatives wrt —Chemical composition and microphysical properties of particles (size distribution, ...) —Non-LTE process parameters (collision rates, production rates, ...) are provided
• Modelling of instrument characteristics (FOV, ILS, ...)
• User-defined numerical accuracy

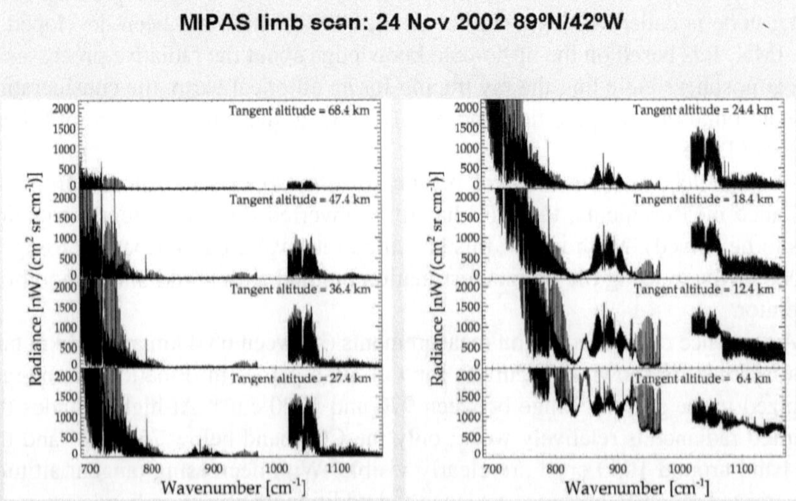

FIGURE 5. Sequence of limb emission spectra in the altitude region between 68.4 km and 6.4 km.

Derivation of Trace Gas Distributions

As mentioned above, from the sequences of measured limb radiance spectra vertical profiles of trace gases can be derived. Combining the vertical profiles along the satellite orbit allows to generate cross sections of atmospheric parameters. In the following examples of results will be shown which prove the enormous capability of the MIPAS experiment.

Figure 6 shows stratospheric cross sections of long-lived trace gases, potential vorticity (PV) and O_3 along the satellite orbit from 40 °S across the South Pole to 40 °S. The three columns are related to different dates: 20 September, 24 September, and 13 October 2002 (from left to right). The time period is connected with the break-up of the Polar Vortex over Antarctica. At 20 September 2004 the Polar Vortex is seen very clearly by the low mixing ratios of N_2O and CH_4 in the lower stratosphere. The potential vorticity (PV) as derived from ECMWF-data confirms this statement. Simultaneously, low ozone values are obviously existing over Antarctica. The split up of the polar vortex is recognized at 24 September 2002 connected with high ozone concentrations at high southern latitudes.

At 13 October 2002 a re-establishment of the polar vortex has happended. On the other hand, this polar vortex is much less pronounced in the middle stratosphere as can be seen in the CH_4 cross section. This example shows clearly the

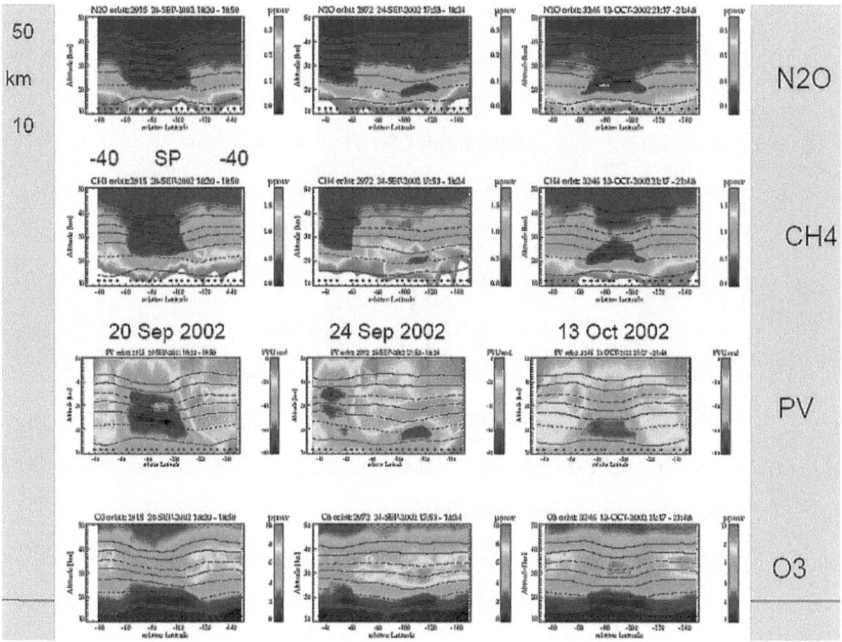

FIGURE 6. Stratospheric cross sections of trace gases and potential vorticity along the satellite orbit from 40 °S across the South Pole to 40 °S for three different dates.

comprehensive information content of the MIPAS data about the stratospheric dynamics [see also Glatthor et al., 2004].

The ozone depletion in the southern polar vortex in spring time can be demonstrated by intercomparing the ozone with the long-lived CFC-11 mixing ratio inside and outside the polar vortex (see Figure 7). At 20 September the O_3 mixing ratio outside the polar vortex for CFC-11 values of 0.02 to 0.01 ppbv is much higher than the corresponding O_3 mixing ratios inside the polar vortex. In case of mid-October the ratio of O_3 mixing ratio outside to inside is even greater than before, i.e., the ozone depletion was ongoing during end of September and beginning of October 2002.

Essential for the stratospheric ozone depletion is the concentration of ClO in the atmosphere; under high $ClONO_2$ concentrations this process is not effective. MIPAS is capable to measure also ClO under disturbed chemical conditions as in the polar winter/spring. Figure 8 shows the stratospheric cross sections of ClO and $ClONO_2$. High amounts of ClO are only visible in the polar lower stratosphere at the day side. $ClONO_2$ shows the typical collar structure over Antarctica [see Höpfner et al., 2004a].

Results on validation of the trace gas distributions have been described in the cited articles. For the main trace gases there is an overview article available based on the ENVISAT validation workshop in May 2004 in Frascati (see http://envisat.esa.int, Fischer H. and H. Oelhaf, ESA-proceedings).

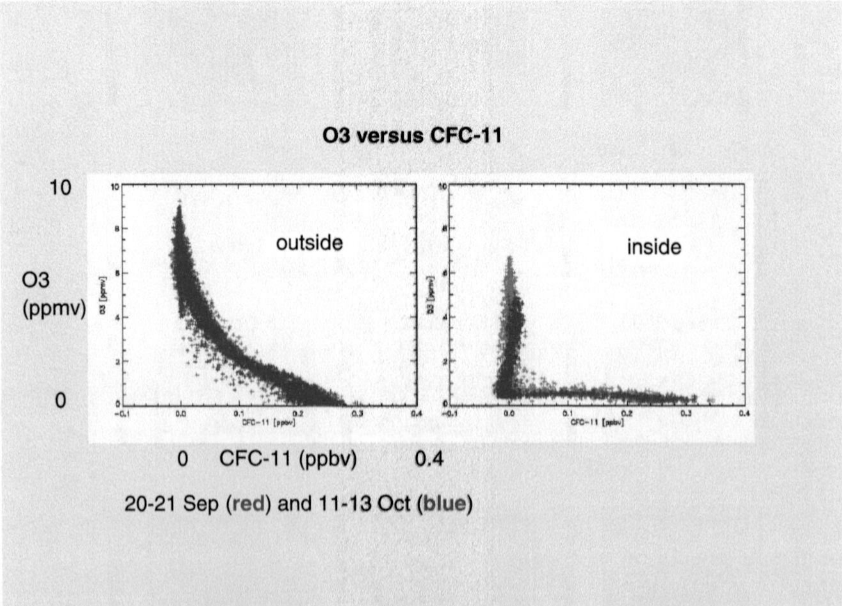

FIGURE 7. Intercomparison of ozone mixing ratios inside and outside the polar vortex for 20/21 September and 11/12/13 October 2002.

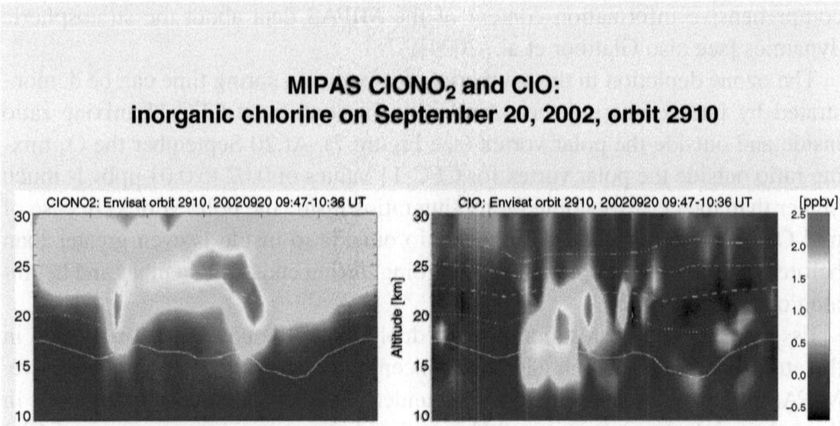

FIGURE 8. Stratospheric cross sections of ClO and $ClONO_2$ for higher latitudes in the southern hemisphere at 20 September 2002.

Determination of Cloud Parameters

MIPAS is a versatile instrument; from the measurements not only trace gas concentrations but also parameters of thin clouds can be derived. Höpfner et al. [2004b] have proven that high resolution broad-band IR spectra can be used to determine the mean size and the volume density of the cloud particles as well as their chemical composition.

The limb spectra of Polar Stratospheric Clouds (PSC) show very special signatures, which are caused by a superimposition of cloud emitted radiance and scattered tropospheric radiance (Figure 9). The interpretation of the spectra as measured along different orbits across the Antarctic region has yielded spatial distributions of PSCs for different dates (see Figure 10). These results show clearly that the development of the PSCs is connected with complex processes in the atmosphere depending on temperature, dynamics, and chemical composition.

Höpfner (personal communication) has already proven that thin cirrus clouds in the tropopause region exhibit the same typical spectral signatures as PSCs, i.e., also size distribution and volume density of these ice particles can be derived.

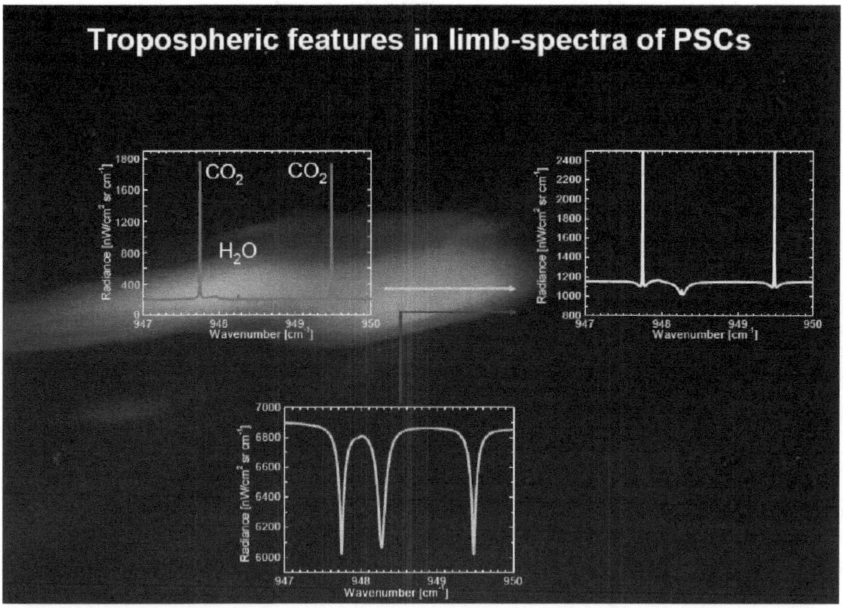

FIGURE 9. Superimposition of PSC emitted radiance and scattered tropospheric radiance for a limb observation.

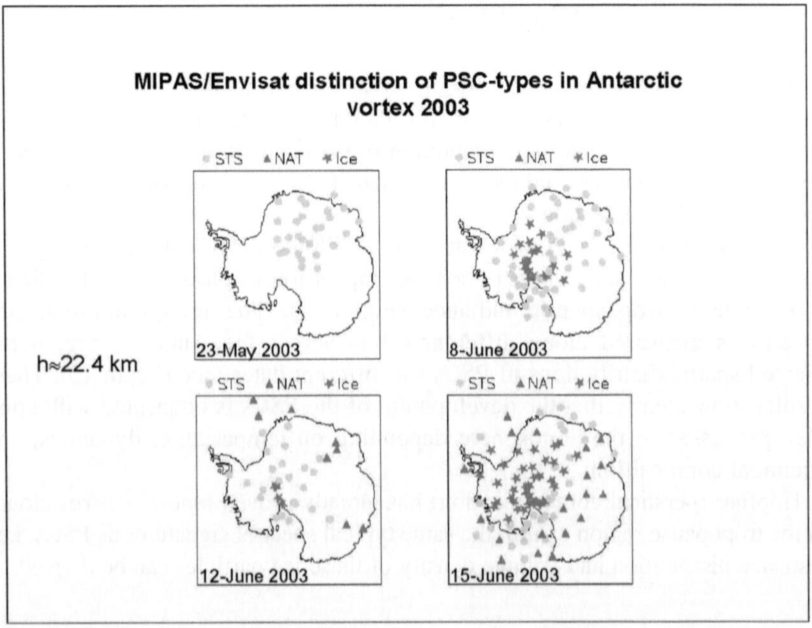

FIGURE 10 Spatial distribution of different types of PSCs over Antarctica at four dates.

Conclusion and Outlook

MIPAS provides high-quality global measurements during day and night of temperature, more than 25 trace constituents and parameters of thin clouds. The validation of the results yields generally good agreement with the measurements of various other experiments. The major warming over the Antarctic stratosphere in September/October 2002 was well covered with MIPAS measurements. First-time ClO concentrations have been detected with IR instruments from space. Simultaneous observations of trace gas concentrations and cloud parameters (PSCs, cirrus clouds) were performed.

Future MIPAS measurements have to be carried out with lower spectral resolution (about 40%) because the interferometer has a problem with the retroreflector slides. On the other hand, the lower spectral resolution does not reduce the information content of the measurements significantly.

Acknowledgement The versatile contributions of several ESA colleagues, members of the MIPAS Scientific Advisory Group, and many scientists of my institute are acknowledged.

References

Fischer, H., (1993) Remote sensing of atmospheric trace gases. *Interdisciplinary Science Reviews* **18**: 185–191.

Fischer, H., C. Blom, H. Oelhaf, B. Carli, M. Carlotti, L. Delbouille, D. Ehhalt, J.-M. Flaud, I. Isaksen, M. Lopez-Puertas, C.T. McElroy, and R. Zander. (2000): Envisat-MIPAS, the Michelson Interferometer for Passive Atmospheric Sounding; An instrument for atmospheric chemistry and climate research, *ESA SP-1229,* C. Readings and R.A. Harris, eds. (European Space Agency, Noordwijk, The Netherlands, 2000).

Glatthor, N., et. al. (2004), Mixing processes during the Antarctic vortex split in September/October 2002 as inferred from source gas and ozone distributions from MIPAS/ENVISAT, *J. Atmos. Sci.*, Special issue on Antarctic Vortex 2002, accepted January 2004.

Höpfner, M., et. al. (2004a), First spaceborne observations of antarctic stratospheric $ClONO_2$ recovery: austral spring 2002, *J. Geophys. Res.*, **109**, D11308.

Höpfner, M. (2004b), Study on the impact of polar stratospheric clouds on high resolution mid-IR limb emission spectra, *J. Q. S. R. T.*, **83**, 93–107.

Rodgers, C.D., (2000), Inverse Methods for Atmospheric Sounding: Theory and Practice, Vol. 2, edited by F.W. Taylor, World Scientific, River Edge, N.J.

Stiller, G.P., (ed.), (2000), The Karlsruhe optimized and precise radiative transfer algorithm (KOPRA), *Wiss. Ber. FZKA 6487*, Forschungszentrum Karlsruhe, Germany.

Part II
Aircraft and Ground-Based Intensive Campaigns

Chapter 6

Probing the Atmosphere with Research Aircraft—European Aircraft Campaigns

ULRICH SCHUMANN

Why Aircraft Campaigns?

Aircraft are used to investigate the atmosphere and the ground by visual observations and by measurements with instruments onboard the aircraft. They allow for

- Targeted measurement of atmospheric properties above ground with in-situ instruments and with airborne remote sensing instruments along a given flight path *(e.g., from 100 m up to 20 km; looking up and down)*
- Measurements at high temporal/spatial resolution *(order 1 s, 10 m, as often required)*
- Data for process studies *(e.g., cloud formation)*
- Data for validation of models *(e.g., chemical transport models)*
- Test and validation of remote sensing (e.g., GOME, SCIAMACHY, MIPAS, e.g., Heland et al., 2002)
- Measurements of otherwise not measurable properties *(e.g., tropospheric NO, turbulence, cloud condensation nuclei)*
- Simultaneous measurements of many parameters *(meteorology, air composition, aerosols, cloud physics etc.)*

This lecture is based on experiences gained within several European research projects with research aircraft, such as POLINAT, EULINOX, INCA, and TROCCINOX. The projects dealt with

— POLINAT: Pollution in the North Atlantic tropopause region induced by aircraft and other sources [Schlager et al., 1997, 1999; Schumann et al., 1995, 2000; Ziereis et al., 1999, 2000],

— EULINOX: Lighting induced nitrogen oxides over Europe [Höller et al., 1999; Huntrieser et al., 2002],

— INCA: Impact of different aerosol and trace gas concentrations and atmospheric dynamics on cirrus clouds in the Southern and Northern hemisphere [Ström et al., 2003; Baehr et al., 2003; Minikin et al., 2003; Ovarlez et al., 2002; Ziereis et al., 2004; Gayet et al., 2002, 2004], and

— TROCCINOX: Nitrogen oxides from lightning (besides cirrus and aerosol properties) near deep convective clouds in a tropical continent [Schumann et al., 2004].

Instead of describing the scientific results of these and related projects, this short contribution concentrates on the European research aircraft fleet and the methodological aspect of airborne measurements.

European Research Aircraft

The European fleet for airborne research is, to a large extent, organized within the EU-network EUFAR; see www.eufar.net. Within this network, the European Commission of the European Union provides financial support for transnational access (preferably for young scientists from countries outside the country of the aircraft operator).

The EUFAR fleet composes presently 22 aircraft within 5 categories. Figure 1 lists the aircraft by type and operator together with a small photo. In addition to these research aircraft, in Europe several instrumented airliners are used, in particular several Airbus A340 aircraft within the project MOZAIC (with measurements of H_2O, O_3, CO, NOy; see paper by J.-P. Cammas in this volume), and a Lufthansa A340-600 (previously LTU B767) carrying a container with a large suite of instruments within the project CARIBIC [Brenninkmeijer et al., 1999]. In 1995–1997, a Swiss Air B747 was equipped with a container carrying instruments to measure NO, NO2, O_3 within the project NOXAR [Brunner et al., 2001]. The Swiss project NOXAR was coordinated with the EU-Project POLINAT 2 [Schumann et al., 2000].

In the near future, it is expected to get access to a new research aircraft known under the project acronym HALO. Similar to the project HIAPER in the USA, HALO will be an instrumented Gulfstream G550. The HALO investment is funded by the German Ministry of Research and Education (BMBF) and by German research institutions (MPG, HGF). HALO will be operated by DLR under an open consortium including in addition the German Science Foundation (DFG), the Leibniz Society (WGL), and other institutes. HALO is expected to become operational in 2008.

Compared to other existing European research aircraft, HALO will go “higher, farther, larger:” up to 15.5 km altitude, more than 8000-km range, and up to 3000-kg payload.

Instrumentation

Some of the research aircraft are equipped with a specific set of instruments, which can only with large efforts be changed into new configuration. However, many of the research aircraft are equipped with devices (racks, openings, inlets, windows, power supply, data recording system, etc.) that allow for equipment

FIGURE 1. The EUFAR fleet.

with variable sets of instruments. For example, the Falcon research aircraft of DLR was equipped with instruments as listed in Table 1, during the field campaign of the project TROCCINOX in Jan–March 2004.

In particular, DLR has developed instruments to be flow on the Falcon for trace gas measurement systems [Schlager et al., 1997], particle instrumentation [Schumann et al., 2002; Minikin et al., 2003], Lidar instruments [Ehret et al., 1999, 2000; Fix et al., 2002] (available for O_3, H_2O, wind profiles), and standard instruments (position, speed, temperature, humidity, wind etc.) [Schumann et al., 1995]. The trace gas measurement system is available to measure many of the components important for photochemical reactions in the troposphere.

FIGURE 1. *(Continued)*

In addition to the particle instrumentation listed in Table 1, an extended set of instruments was available within the project INCA, to cover the whole size range from a few nanometers to millimeters; see Figure 2.

During TROCCINOX, the Lidar was flown looking upwards to measure Lidar backscattering—at wavelengths 532 and 1064 nm—and H_2O by differential absorption with two wavelengths near 935 nm, in the altitude range up to 16 km; see Figure 3 for an example.

During transit from Germany to Brazil and back, the Lidar was looking downward to measure aerosol and H_2O profiles from 11 km downward to the ground within the troposphere. In other projects, the Lidar systems were flown to measure O_3 and wind profiles.

TABLE 1. Falcon Instrumentation for TROCCINOX

Species/Parameter	Technique	Δt	Precision/Accuracy
Aerosol Profile	DIAL	5 sec	1/5%
Remote H_2O Profile	DIAL upward or downward	1 min	5/10%
Ozone	UV absorption	5 sec	1/5%
NO	Chemiluminescence	1 sec	3%/10%
NO_y	Chemiluminescence + Au converter	1 sec	5%/15%
CO	VUV fluorescence	5 sec	1/3 nmol/mol
Condensation nuclei (CN)	Condensation particle size analyzer (lower cutoff 4, 9, 13 nm)	1 sec	5/10%
Nonvolatile CN	CN counter with heated inlet (lower cutoff 13 and 50 nm)	1 sec	5/10%
Aerosol absorption coefficient	PSAP (particle soot absorption photometer)	30 sec (BL)	10/20%
Aerosol and cloud element size distribution	PCASP-100X (0.2–1.0 μm) FSSP-300/300 (0.3–20/5–100 μm)	20–60 sec	10/20%
Position, wind	INS, GPS, 5-hole probe	1 sec	200 m (horiz.) 0.05/1 m/s (h) 0.05/0.3 m/s (v)
Temperature	Rosemount	1 sec	0.1/0.5 K
Humidity	Lyman-alpha	1 sec	0.01/0.3 g/m^3
NO_2 photolysis, $J(NO_2)$	Filter radiometer	1 sec	1 E-4/5 E-4

The spatial and temporal resolution of airborne measurements is usually far higher than what can be achieved by remote sensing -on the other hand, the coverage reached by satellite instruments is far higher than for aircraft. For properties influenced by turbulence or clouds in the atmosphere and for short lived chemical species, high resolution is particular important. Typically the instruments reach a temporal resolution of about 1 s. Other important properties of instruments are precision and accuracy, i.e., the relative and absolute resolution of the measured parameter. See Table 1 for the properties of the Falcon instrumentation during TROCCINOX.

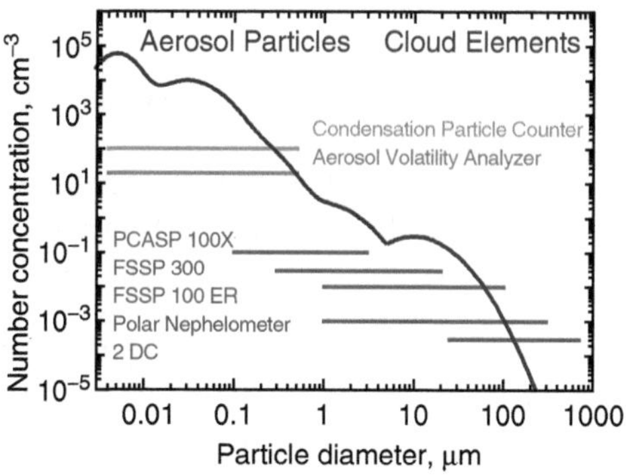

FIGURE 2. Typical aerosol size spectrum and its coverage by various instruments. [A. Petzold, A, Minikin, DLR, pers. Communication, 2003.]

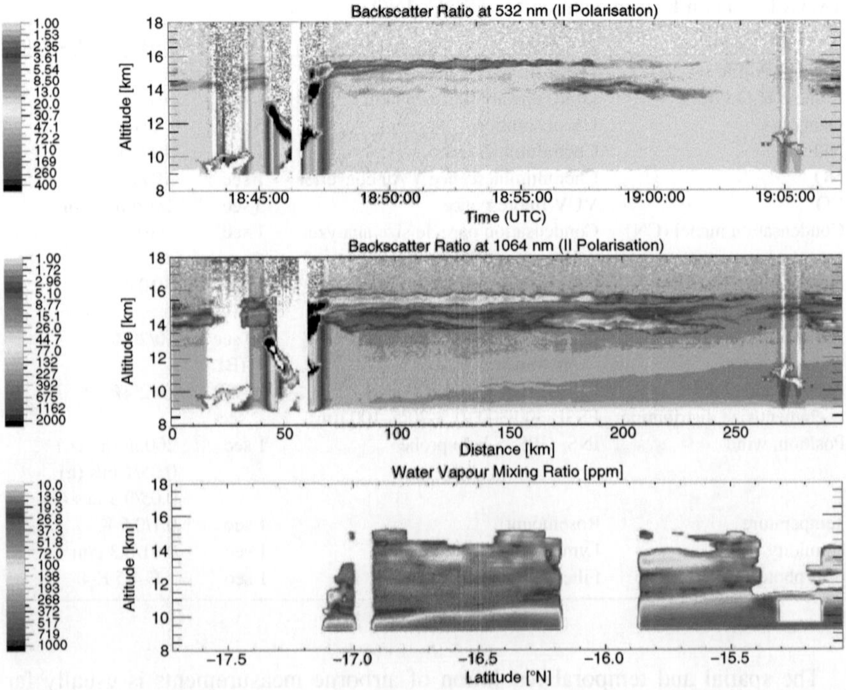

FIGURE 3. Observations of backscatter and water vapor near the anvil of a tropical thunderstorm during TROCCINOX [G. Ehret, A. Fix, H. Flentje, M. Wirth, personal communication, 2004].

Quality Checks

An important task in performing airborne measurements are quality checks of the instrumentation. Such checks include many activities such as model analysis and laboratory studies, preflight, inflight, and postflight tests and calibrations. One specific activity of large importance are intercomparison flights, where two aircraft fly close to each other, preferably wing by wing, to measure the same atmospheric parameter with their individual instruments.

For example, during the summer 2000 Export aircraft campaign (European eXport of Precursors and Ozone by long-Range Transport), see Figure 4, two comprehensively instrumented research aircraft, the British C-130 and the German Falcon, measuring a variety of chemical species flew wing tip to wing tip for a period of one and a quarter hours [Brough et al., 2003]. During this interval a comparison was undertaken of the measurements of (NO), NO_y, CO, and O_3. The comparison was performed at two different flight levels, which provided a 10-fold variation in the concentrations of both NO (10 to 1000 pmol/mol) and NO_y (200 to over 2500 pmol/mol). Large peaks of NO and NO_y observed from the *Falcon 20*, which were at first thought to be from the exhaust of the *C-130*,

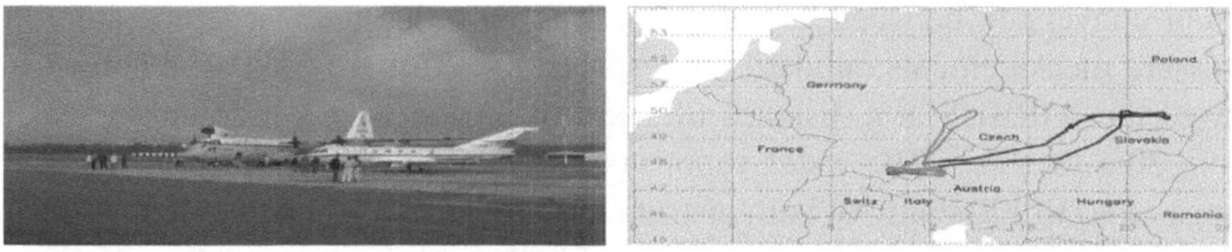

FIGURE 4. Left. The MRF *C-130* and the DLR *Falcon 20* during EXPORT. Right: Flight paths of the *Falcon* (light gray) and the *C-130* (dark gray) [Brough et al., 2003].

were also detected on the 4-channel $NO_{x,y}$ instrument aboard the *C-130.* See Figure 5, for example. These peaks were a good indication that both aircraft were in the same air mass and that the *Falcon 20* was not in the exhaust plume of the *C-130.* Correlations and statistical analysis are presented between the instruments used on the two separate aircraft platforms. These were found to be in good agreement, giving a high degree of correlation for the ambient air studied. Any deviations from the correlations are accounted for in the estimated inaccuracies of the instruments. These results help to establish that the instruments aboard the separate aircraft are reliably able to measure the corresponding chemical species in the range of conditions sampled.

An intercomparison of airborne in-situ water vapor measurements by two European research projects (MOZAIC and POLINAT) was performed from aboard the respective Airbus (MOZAIC) and Falcon (POLINAT) aircraft [Helten et al., 1999]. The intercomparison took place southwest of Ireland on September 24, 1997, at 239 hPa flight level. MOZAIC uses individually calibrated capacitive humidity sensors for the humidity measurement. POLINAT employs a cryogenic frost-point hygrometer developed for such measurements [Ovarlez et al., 2003].

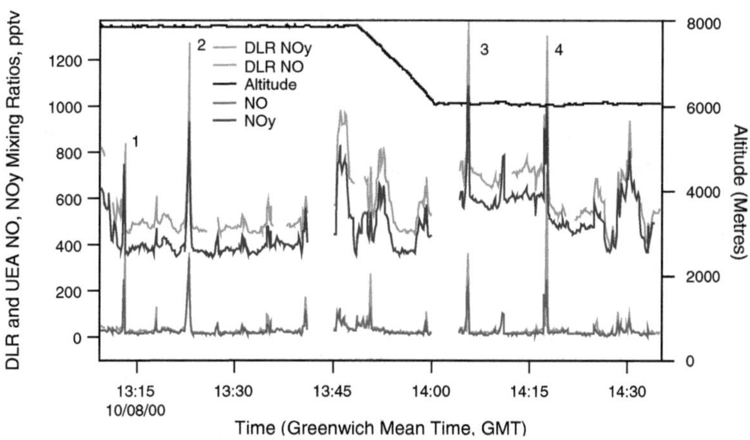

FIGURE 5. The mixing ratios for NO and NO_y observed at altitudes between 6000 and 8000 m, during the comparison over southern Germany [Brough et al., 2003].

For conversion between humidity and mixing ratio, ambient temperature and pressure measurements onboard the respective aircraft are used. The Falcon followed the AIRBUS at a distance of 7 km to 35 km with a time-lag increasing from 30 s to 160 s. The water vapor volume mixing ratio measurements in the range of 80 to 120 ppmv of both instruments are in excellent agreement, differing by less than ±5%, where the trajectories of both aircraft are very close. However, the relative humidity (RH) calculated from POLINAT frost-point measurements and the Falcon -PT500 temperature sensor is up to 15% higher relative to the RH of MOZAIC. The agreement improved to within 5% when using the temperature measurement of the PT100 sensor instead of the temperature measurement of the PT500 sensor for RH determination of POLINAT.

Campaign Planning

An important part of any airborne campaign is the campaign planning. This topic is usually not reported in the scientific literature. It simply belongs to the tasks of any such experiment. Nevertheless, it is important to recognize that this task usually takes months before the experiment, and its quality is one of the main reason for project success or failure. The "White Book" by Rob MacKenzie et al. (written as part of the TROCCINOX project), includes the following tasks:

Pre-Campaign Phase: Instrument preparation and upgrading, model development, tools development for flight planning, and logistics and operational planning, aircraft preparation.

Campaign Phase: Calibration of the instruments, campaign performance, data analysis

Interpretation Phase: Modeling of the observed scenarios using state-of-the-art models and hypotheses tests by intercomparison with the data sets.

The results of the Pre-Campaign Planning are documented in a document called "White Book". Its content includes

— Key scientific questions
— Campaign methodology: Site selection, observational platforms, alternate airports, flight objectives, instrumentation
— Flight templates
— Timelines
— Contingency
— Management
— Briefing

Some of the ingredients of the White Book were presented in the workshop. The reader may refer to the project homepage for further details. An essential part of campaign organizations are recalculations of the chemical composition of the atmosphere, see Flatoy et al. [2000].

Databases

Usually, the results of the data of a measurement campaign are collected in a project specific database, which includes all the original measurements for access by all project partners. Usually, the data banks are pass-word-controlled until at least one year after the project ends. It is common practice to allow other users access on the condition that the results are published in agreement with the experimenters, including co-authorship were reasonable.

The databases can usually be found via the homepages of the specific projects in the internet, e.g., http://www.pa.op.dlr.de/polinat/,../inca/,../eulinox/,./troccinox/, etc.

Comparison to Models

Collection of data from various projects for a class of measurements is required if one wants to compare various chemical transport models or other models to the measured results available. Some well-known examples of such data collection activities are described by Emmons et al. [2000]. The "classical" approach for evaluating a model in such studies was to aggregate the observations over specific domains, altitude ranges, and time periods, and then to compare statistical quantities such as mean or median values and standard deviations for these aggregates with corresponding model data.

However, the model fields were usually not sampled at exactly the same times and positions as the measurements but rather averages over entire time periods (e.g., monthly means) and domains (e.g., over a range of grid cells) were calculated because such quantities can easily be derived from standard model output. A more direct approach was used, e.g., by Brunner et al. [2003], compares each measured data point with its temporally and spatially interpolated model counterpart. Such a "point-by-point" analysis requires us to simulate exactly the same time periods when the measurements were obtained.

Within the TRADEOFF project [Brunner et al., 2003], an extensive database including all major research aircraft and commercial airliner measurements between 1995 and 1998 as well as ozone soundings were established. The database is constructed to support direct comparison between model results and measurements. Moreover, quantitative methods were applied to judge model performance including the calculation of average concentration biases and the visualization of correlations and RMS errors in the form of so-called Taylor diagrams.

The comparisons show quite good agreements for tracers that are sufficiently smooth (such as ozone) but often rather bad agreements for short-lived trace gases with ill-defined sources and sinks (such as NO). In particular, the NO measurements show a larger dynamic range than the models. The correlation coefficient between modeled and measured results, see Figure 6 for example, is often far below 0.5. This indicates that still a lot of development work and higher numerical model resolution is required to bring chemical transport models into a state of satisfying accuracy.

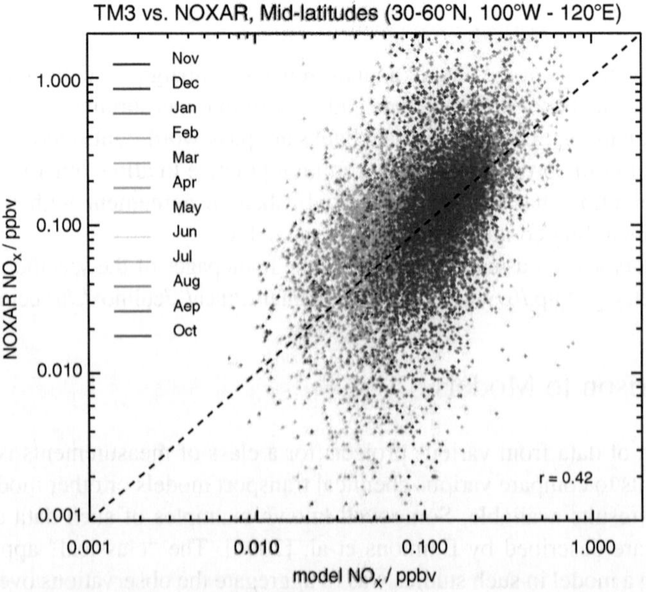

FIGURE 6. Scatter plot of model results versus NOXAR NO [Brunner et al., 2003].

Conclusions

Research aircraft form an essential component of the atmospheric (and Earth observation) research infrastructure. Usage of aircraft requires skills, as indicated in this lecture, not only with respect to the underlying science, but also with respect to aviation, instrumentation, campaign management, and data evaluation.

References

Baehr, J., H. Schlager, H. Ziereis, P. Stock, P. van Velthoven, R. Busen, J. Ström, U. Schumann, Aircraft observations of NO, NO_y, CO, and O_3 in the upper troposphere from 60 °N to 60 °S—Interhemispheric differences at mitlatitudes, *Geophys. Res. Lett.* **30**(1), 10.1029/2003GL016935, 2003.

Brenninkmeijer, C.A.M., Crutzen, P.J., Fischer, H., Güsten, H., Hans, W., Heinrich, G., Heintzenberg, J., Hermann, M., Immelmann. T., Kersting, D., Maiss, M., Nolle, M., Pitscheider, A., Pohlkamp, H., Scharffe, D., Specht, K., and Wiedensohler, A. (1999) CARIBIC—civil aircraft for global measurement of trace gases and aerosols in the tropopause region, *J. Atmos. Oceanic Technol.* **16**, 1373–1383, 1999.

Brough, N., C. Reeves, S.A. Penkett, K. Dewey, J. Kent, H. Barjat, P. S. Monks, H. Ziereis, P. Stock, H. Huntrieser and H. Schlager, Intercomparison of aircraft instruments on board the *C-130* and *Falcon 20* over southern Germany during EXPORT 2000, ACP, 2003.

Brunner, B., J. Staehelin, D. Jeker, H. Wernli, and U. Schumann, Nitrogen oxides and ozone in the tropopause region of the Northern Hemisphere: Measurements from commercial aircraft in 1995/96 and 1997. *J. Geophys. Res.,* **106**, 27673–27699, 2001.

Brunner, D., Staehlin, J., Rogers, H.L., Köhler, M., Pyle, J., Hauglustaine, L., Jourdain, T., Berntsen, K., Gauss, M., Isaksen, I., Meijer, E., Velthoven van, P., Pitari, G., Mancini, E., Grewe, V., Sausen, R., 2003: An evaluation of the performance of chemistry transport models by comparison with research aircraft observations. Part 1: Concepts and overall model performance. *Atmos. Chem. Phys.*, **3**, 1609–1631, 2003.

Ehret, G., K.P. Hoinka, J. Stein, A. Fix, C. Kiemle, G. Poberaj, Low-stratospheric water vapor measured by an airborne DIAL. *J. Geophys. Res.*, **104**, D24, 31,351–31,359, 1999.

Ehret, G., H.H. Klingenberg, U. Hefter, A. Assion, A. Fix, G.S. Poberaj, S. Berger, S. Geiger, Q. Lü, High peak and average power all-solid-state laser systems for airborne LIDAR applications. *LaserOpto*, **32**(1), 29–37, 2000.

Emmons, L. K., D.A. Hauglustaine, J.-F. Müller, M.A. Carroll, G.P. Brasseur, D. Brunner, J. Staehelin, V. Thouret, A. Marenco, Data composites of airborne observations of tropospheric ozone and its precursors, *J. Geophys. Res.*, **105**, (20), 497–538, 2000.

Fix, A., M. Wirth, A. Meister, G. Ehret, M. Pesch, D. Weidauer, Tunable ultraviolett optical parametric oscillator for differential absorption Lidar measurements of tropospheric ozone. *Appl. Phys. B*, **75**, (10.1007/s00340-002-0964-y), 153–163, 2002.

Flatoy, F., O. Hov, H. Schlager, Chemical forecasts used for measurement flight planning during POLINAT 2. *Geophys. Res. Lett.*, **27**(7), 951–954, 2000.

Gayet, J.-F., F. Auriol, A. Minikin, J. Strom, M. Seifert, R. Krejci, A. Petzold, G. Febvre and U. Schumann, Quantitative measurements of the microphysical and optical properties of cirrus clouds with four different in situ probes: Evidence of small ice crystals, *Geophys. Res. Lett.*, **29**(24), 2230, doi:10.1029/2001GL014342, 2002.

Gayet, J.-F., J. Ovarlez, V. Shcherbakov, J. Ström, U. Schumann, A. Minikin, F. Auriol, A. Petzold, M. Monier, Cirrus cloud microphysical and optical properties at southern and northern midlatitudes during the INCA experiment. *J. Geophys. Res.*, **109**, D20206, doi:10.1029/ 2004JD004803, 2004.

Heland, J., H. Schlager, A. Richter, J.P. Burrows, First comparison of tropospheric NO_2 column densities retrieved from GOME and in situ aircraft profile measurements. *Geophys. Res. Lett.*, **29**, 20 (10.1029/2002GL015528), 44-1-44-4, 2002.

Helten, M., H.G.J. Smit, D. Kley, J. Ovarlez, H. Schlager, R. Baumann, U. Schumann, P. Nedelec, A. Marenco, In-flight comparison of MOZAIC and POLINAT water vapor measurements. *J. Geophys. Res.*, **104**, D21, 1999, 26,087-26,096.

Höller, H., U. Finke, H. Huntrieser, M. Hagen, and C. Feigl, Lightning produced NO_X (LINOX)—Experimental design and case study results, *J. Geophys. Res.*, **104**, 13911–13922, 1999.

Huntrieser, H., Ch. Feigl, H. Schlager, F. Schröder, Ch. Gerbig, P. van Velthoven, F. Flatøy, C. Théry, A. Petzold, H. Höller, U. Schumann, Airborne measurements of NOx, tracer species and small particles during the European Lightning Nitrogen Oxides Experiment, *J. Geophys. Res.*, **107**(11), AAC 2-1–AAC 2-27, 10.1029/2000JD000209, 2002.

Minikin, A., A. Petzold, J. Ström, R. Krejci, M. Seifert, P. van Velthoven, H. Schlager, U. Schumann, Aircraft observations of the upper tropospheric fine particle aerosol in the Northern and Southern Hemispheres at midlatitudes, *Geophys. Res. Lett.* **30**(10), 10.1029/2002GL016458, 2003.

Ovarlez, J., J.-F. Gayet, K. Gierens, J. Ström, H. Ovarlez, F. Auriol, R. Busen, U. Schumann, Water vapour measurements inside cirrus clouds in Northern and Southern hemispheres during INCA, *Geophys. Res. Lett.*, **29**(16), 60-1–60-4, 200210. 1029/2001 GL014440, 2002.

Schlager, H., P. Konopka, P. Schulte, U. Schumann, H. Ziereis, F. Arnold, M. Klemm, D., E. Hagen, P. D. Whitefield, J. Ovarlez, In situ observations of airtraffic emission signatures in the North Atlantic flight corridor. *J. Geophys. Res.*, **102**, 10,739–750, 1997.

Schlager, H., P. Schulte, F. Flatoy, F. Slemr, P. van Velthoven, H. Ziereis, and U. Schumann, Regional nitric oxides enhancements in the North Atlantic flight corridor observed and modeled during POLINAT 2—a case study, *Geophys. Res. Lett.*, **26**, 3061–3064, 1996.

Schumann, U., P. Konopka, R. Baumann, R. Busen, T. Gerz, H. Schlager, P. Schulte, H. Volkert: Estimate of diffusion parameters of aircraft exhaust plumes near the tropopause from nitric oxide and turbulence measurements. *J. Geophys. Res.*, **100**, 14147–14162, 1995.

Schumann, U., H. Schlager, F. Arnold, J. Ovarlez, H. Kelder, Ø. Hov, G. Hayman, I. S. A. Isaksen, J. Staehelin, and P. D. Whitefield, Pollution from aircraft emissions in the North Atlantic flight corridor: Overview on the POLINAT projects, *J. Geophys. Res.*, **105**, 3605–3631, 2005.

Schumann, U., F. Arnold, R. Busen, J. Curtius, B. Kärcher, A. Kiendler, A. Petzold, H. Schlager, F. Schröder, K.-H. Wohlfrom, Influence of fuel sulfur on the composition of aircraft exhaust plumes: The experiments SULFUR 1–7, *J. Geophys. Res.*, **107**(D15), 10.1029/2001JD000813, AAC 2-1 – AAC 2-27, 2002.

Schumann, U., H. Huntrieser, H. Schlager, L. Bugliaro, C. Gatzen, H. Hoeller, Nitrogen Oxides from thunderstorms—Results from experiments over Europe and the Continental tropics, DACH—Deutsch-Österreichisch-Schweizerische Meteorologen-Tagung, 7.–10. September 2004, Karlsruhe, Deutschland, Proceedings in press (2004), http://imk-msa.fzk.de/dach2004/

Ström, J., M. Seifert, J. Ovarlez, A. Minikin, J.-F. Gayet, R. Krejci, A. Petzold, F. Auriol, R. Busen, U. Schumann, B. Kärcher, W. Haag and H.-C. Hansson, Cirrus cloud occurrence as function of ambient relative humidity: A comparison of observations obtained during the INCA experiment, *Atmos. Chem. Phys.*, **3**, 1807–1816, 2003.

Ziereis, H., H. Schlager, P. Schulte, I. Köhler, R. Marquardt, C. Feigl, In situ measurements of the NO_x distribution and variability over the Eastern North Atlantic. *J. Geophys. Res.*, **104**, 16021–16032, 1999.

Ziereis, H., H. Schlager, P. Schulte, P. E. J. van Velthoven, F. Slemr, Distributions of NO, NO_x and NO_y in the upper troposphere and lower stratosphere between 28° and 61° during POLINAT 2. *J. Geophys. Res.*, **105**, D3, 3653–3664, 2000.

Ziereis, H., A. Minikin, H. Schlager, J. F. Gayet, F. Auriol, P. Stock, J. Baehr, A. Petzold, U. Schumann, A. Weinheimer, B. Ridley, J. Ström, Uptake of reactive nitrogen on cirrus cloud particles during INCA. *Geophys. Res. Lett.*, **31**(5), L05115 10.1029/2003GL018794, 2004.

Chapter 7

MOZAIC—Measuring Tropospheric Constituents from Commercial Aircraft

JEAN-PIERRE CAMMAS

Introduction

The overall objective of MOZAIC is to improve our knowledge on the concentrations and life cycles of ozone, water vapor, and related trace gases (CO and NO_y) in the atmosphere. Particular emphasis is on the upper troposphere and lower stratosphere (UT/LS), a region where large gaps exist in our present knowledge and which is of great importance for climate.

Three phases of MOZAIC have been funded by the European Commission from 1993 to February 2004. In the first phase of MOZAIC (1994–1996) automatic devices for measurements of ozone and water vapor have been installed on 5 Airbus A340. Lufthansa, Air France, Austrian, and Sabena fly the MOZAIC instrumentation free of charge. In the second phase of MOZAIC (1997–2000), O_3 and H_2O measurements were continued while new devices for measurements of carbon monoxide (CO) and nitrogen oxides (NO_y) were built. In the third phase of MOZAIC (2001–2004), new devices for CO were installed onboard the five aircraft and the NO_y instrument was installed onboard one aircraft. From August 1994 to December 2003, 22,003 flights (156,652 flights hours) have been performed by the aircrafts at a quasi-global scale.

The MOZAIC consortium reflects a unique collaboration between scientists (CNRS France, Forschungszentrum Germany, Météo-France, University of Cambridge UK), the aircraft manufacturer Airbus, the airlines (Lufthansa, Air France, and Austrian) that transport free of charge the MOZAIC instrumentation, and the European Commission.

Instrumentation

The MOZAIC ozone instrument is an improved version of a commercial dual-beam UV absorption instrument [Marenco et al., 1998; Thouret et al., 1998]. The accuracy is ±2 ppbv, the precision ±2%, and the response time 4 sec. The instrument is replaced every C-check of the aircraft (about 6 months) and is recalibrated at the laboratory.

The MOZAIC relative humidity instrument is a capacitive RH sensor (Humicap-H) mounted in a Rosemount Probe System (Helten et al., 1998). The accuracy is less than 10%, the precision is about ±7%, the response time depends on temperature (a few secondes at ground to 2 minutes at cruise altitude). Sensor replacement and pre- and postcalibration are made in laboratory every C-check of the aircraft.

The MOZAIC CO instrument is an improved version of a commercial IR gas filter correlation model. Its accuracy is ±5 ppbv, precision ±5%, detection limit 10 ppbv, response time 30 sec [Nédélec et al., 2003]. Sensor replacement and pre- and postcalibration are made in laboratory every C-check of aicraft.

The aeronautical certification for the instruments above has been delivered by Airbus.

The MOZAIC NO_y instrument detects NO by chemiluminescence with O_3 in combination with catalytic conversion of the other NO_y compounds to NO at 300 °C on a gold surface in the presence of H_2 [Volz-Thomas et al., 2004]. Sensitivity of 0.3–0.5 cps/ppt gives a detection limit of better than 30–50 ppt for an integration time of 4 s, and 150–300 ppt at the maximum resolution of the instrument (10 Hz). It is designed for unattended operation during 400–800 flight hours. The total weight is 50 kg, including calibration system, compressed gases, mounting, and safety measures. Aeronautical certification is delivered by Lufthansa technics.

Database

The MOZAIC database is located in Toulouse at CNRM (Centre National de la Recherche Météorologique, Météo-France). It contains raw data (full resolution) and averaged data (lower resolution for the purpose of comparison with results from large-scale models). It also contains other relative meteorological data, like the difference of pressure between the aircraft and the dynamical tropopause on ECMWF analyses (taken at 2 units of potential vorticity), or like quick views of satellite images.

The access to MOZAIC data is public and subject to the signature of a protocol by co-investigators. Details of the protocol are provided at the end. About 50 international groups are using MOZAIC data.

Main Results

Scientific achievements:

1. Climatology of ozone at a quasi-global scale (a) in the UTLS at mid-latitudes showing (i) a spring maximum in the LS and a spring-to-summer maximum in the UT, (ii) a large interannual variability with a strong positive anomaly

in 1998–1999 from Northeast America to Europe, (b) on vertical profiles showing (i) the 1998–1999 anomaly over Paris, Frankfurt, and New York, (ii) ozone maxima confined in the lower troposphere over Africa during the dry season in relation to biomass burning and photochemistry.

2. Climatology of relative humidity showing that the UT is considerably moister than previously assumed. Over the North Atlantic, more than 30% of the data show substantial ice super-saturation (ISS). This finding is not reflected in the ECMWF analyses, which significantly underestimate the measurements above a relative humidity of 50%. MOZAIC shows that ISS is a small-scale phenomenon and that the distribution of humidity in the tropical UT is bi-modal.
3. First quasi-global scale distributions of CO and NO_y in the UTLS region showing the strong impact of small-scale convection and large-scale uplifting by warm conveyor belts on the composition of the UT, especially over the east coast, of North America and over Asia.
4. Quantitative assessments of stratosphere-troposphere exchanges (subtropical filaments associated with Rossby wave breaking, mid-latitude tropopause folds, ...) using a variety of Lagrangian methodologies applied to numerical weather analyses and using mesoscale and global models.
5. Assessment of present modeling capabilities for the UTLS region by Chemistry Transport Models through validation studies against MOZAIC data, showing a relatively good ability of models tested (TOMCAT, Cambridge; MOCAGE, Météo-France, MOZART-2, MPI-Hamburg) to reproduce the ozone cycle in the UTLS and identifying some deficiencies in the vertical transport and in the parameterizations of clouds and chemistry. Numerical assessments by these models of the perturbation of the UTLS region due to aircraft emissions fall within the most recent values published in the literature.
6. Detection of recent aircraft emissions with 10-Hz measurements of the NO_y device. A climatological survey shows that the such plumes comprise about 1.5% of the total NO_y over the North Atlantic flight corridor, USA, and Europe, without significant spatial or seasonal variations. Smaller fractions are observed over North Africa and much higher fractions occur over East Asia, most likely because of narrower corridors in this region.

Main deliverables:

1. Ten years of quality-assured measurements of ozone (O_3) and water vapor (H_2O); more than 2 years of quality-assured measurements of carbon monoxide (CO), 3 years of total odd-nitrogen (NO_y); from 5 commercial aircraft between 0 and 12 km altitude over the period 1994–2004 (> 22,000 intercontinental flights). The database is open to public access.
2. Automatic instruments with operating procedures and data retrieval algorithms for routine measurements of O_3, H_2O, CO, and NO_y aboard in-service aircraft.
3. Improved knowledge of the climatology and interannual variability of O_3 in the UTLS region.

4. Improved knowledge of the distribution of water vapor in the UT, on the quality of its representation by ECMWF analyses, on the ice super-saturation phenomena, and on processes responsible for its distribution in the tropical UT.
5. First large-scale distributions of CO and NO_y in the UTLS region, including seasonality, and of vertical profiles at quasi-global scale.
6. Improved knowledge of the impact of (i) stratospheric intrusions on the composition of the troposphere, and (ii) of vertical transport from the boundary layer on the chemical composition of the UT region.
7. Improvement of our capacity in validating Chemistry Transport Models, which are the tools used for quantification of the UTLS ozone budget, for assessment of climate change and impact of aircraft, and for providing answers to policy makers and industry.

Future of Mozaic

A memorandum of understanding has been signed between Airbus, Lufthansa, Air France, Austrian, and Centre National de la Recherche Scientifique to continue the present MOZAIC phase until December 2007. The internationally expressed need for representative high-quality in-situ data for the UTLS and the technical expertise developed in MOZAIC lead to an initiative for extending the MOZAIC concept into a large European infrastructure that will provide real-time in-situ measurements of trace gases in the UTLS region to the European GMES initiative. As a first step, a design study for new infrastructures (IAGOS: Integration of routine aircraft measurements into the Global Observing System) submitted in the last call of FP6 for the design and certification of new instrumentation has been accepted by the European Commission and has started in April 2005 for a 3-year duration.

Data Access Protocol

Each MOZAIC co-investigator is asked to sign the MOZAIC data protocol (http://www.aero.obs-mip.fr/mozaic/) and to provide piece of information on:

- The research that will be performed in the next two years with use of MOZAIC data (1 page maximum),
- The MOZAIC Principal Investigator(s) to which collaboration or/and co-authorship could be proposed.

Principal investigators of the MOZAIC program:

J.-P. Cammas (LA/CNRS, Toulouse, France, MOZAIC-III coordinator, camjp@aero.obs-mip.fr)
V. Thouret (LA/CNRS, Toulouse, France, thov@aero.obs-mip.fr)
H. Smit (FZJ, Jüliech, Germany, h.smit@fz-juelich.de)
F. Karcher (Météo-France, Toulouse, France, Fernand.Karcher@meteo.fr)
A. Volz-Thomas (FZJ, Jüliech, Germany, a.volz-thomas@fz-juelich.de)

References

Asman, W.A.H., M.G. Lawrence, C.A.M. Brenninkmeijer, P.J. Crutzen, J.W.M. Cuijpers, P. Nédélec, Rarity of upper-tropospheric low O_3 mixing ratio events during MOZAIC flights. *Atmos. Chem. Phys.* **3**, 1541–1549, 2003.

Baray J.L., T. Randriambelo, S. Baldy, G. Ancellet, Comment on 'Tropospheric O_3 distribution over the Indian Ocean during spring 1995 evaluated with a chemistry-climate model', by A.T.J. de Laat et al., *Journal of Geophysical Research*, **106**(D1), 1365–1368, 2001.

Baray J.L., S. Baldy, R.D., Diab, J.P. Cammas, Dynamical study of a tropical cut-off low over South Africa, and its impact on tropospheric ozone, *Atmospheric Environment*, **37**(11), 1475–1488, 2003.

Borchi F., A. Marenco, Discrimination of air masses near the extratropical tropopause by multivariate analyses from MOZAIC data. *Atmos. Environ.* **7**, 1123–1135, Mar. 2002.

Bregman A., M.C. Krol, H. Teyssedre, W.A. Norton, A. Iwi, M. Chipperfield, G. Pitari, J.K. Sundet, J. Lelieveld, Chemistry-transport model comparison with ozone observations in the midlatitude lowermost stratosphere. *J. of Geophys. Res.*, **106**(D15), 17479–17496, 2001.

Bregman, B., A. Segers, M. Krol, E. Meijer and P. van Velthoven: On the use of mass-conserving wind fields in chemistry-transport models. *Atmos. Chem. Phys.* **3**, 447–457, 2003.

Brioude et al., MOZAIC case study of stratosphere-troposphere exchanges in a summer extratropical low, To be submitted to *Atmos. Chem. Phys Discuss.*, 2005.

Cammas J.-P., S. Jakobi-Koali, K. Suhre, R. Rosset, A. Marenco, Atlantic subtropical potential vorticity barrier as seen by MOZAIC flights. *J. of Geophys. Res.*, **103**, 25681–25693, 1998.

Cathala, M.L., J. Pailleux, et V.-H. Peuch, Improving global simulations of UTLS ozone with assimilation of MOZAIC data. *Tellus*, **55B**, 1–10, 2003.

Cho, J.Y.N., V. Thouret, R.E. Newell, A. Marenco, Isentropic scaling analysis of ozone in the upper troposphere and lower stratosphere, *J. Geophys. Res.*, **106**, 10,023–10,038, 2000.

Cho, J.Y.N., E. Lindborg, Horizontal velocity structure functions in the upper troposphere and lower stratosphere: 1. Observations, *J. Geophys. Res.*, **106**, 10,223–10,232, 2001.

Cooper O.R., A. Stohl, S. Eckhardt, D. Parrish, S.J. Oltmans, B.J. Johnson, P. Nedelec, F.J. Schmidlin: A comparison of the east and west coasts of the United States during spring 2002: Different pollutant transport pathways but nearly identical tropospheric ozone burden. Submitted to JGR.

Crowther, R.A., K.S. Law, J.A. Pyle, S. Bekki, H.G.J. Smit, Characterising the effect of large-scale model resolution upon calculated OH production using MOZAIC data, *Geophys. Res. Lett.*, DOI 10.1029/2002GL014660, 2002.

Crowther, R.A., K.S. Law, J.A. Pyle, A. Volz-Thomas, W. Paetz, P. Nedelec, J.P. Cammas, Investigating the origin of O_3 in the upper troposphere using MOZAIC data, to be submitted to *Atmos. Chem. Phys.*, 2005.

Crowther, R.A., K.S. Law, J.A. Pyle, H.G.J. Smit, P. Nedelec, Evaluation of modelled photochemical OH production using MOZAIC data, to be submitted to *Atmos. Chem. Phys.*, 2005.

Dethof A, A. O'Neill, J.M. Slingo, et al., A mechanism for moistening the lower stratosphere involving the Asian summer monsoon. *Quart. J. of the Roy. Met. Soc.*, **125**(556): 1079–1106, 1999.

Diab R., A. Raghunandan, A. Thompson, V. Thouret, Classification of tropospheric ozone profiles based on MOZAIC aircraft data. *Atmos. Chem. and Phys.*, **3**, 713–723, 12-6-2003.

Edwards, D.P., J.-F. Lamarque, J.-L. Atti\'e, L.K. Emmons, A. Richter, J.-P. Cammas, J.C. Gille, G.L. Francis, M.N. Deeter, J. Warner, D. Ziskin, L.V. Lyjak, J. R. Drummond, J. P. Burrows, Tropospheric ozone over the tropical Atlantic: A satellite perspective, *J. Geophys. Res.* **108**, 4237, doi:10.1029/2002JD002927, 2003.

Emmons, L.K., D.A. Hauglustaine, J-F. Muller, M.A. Caroll, G.P. Brasseur, D. Brunner, J. Staehelin, V. Thouret, A. Marenco, Data composites of airborne observations of tropospheric ozone and its precursors, *J. Geophys. Res.*, **105**, 20,497–20,538, 2000.

Forster, C., A. Stohl, P. James, V. Thouret, The residence times of aircraft emissions in the stratosphere using a mean emissions inventory and emissions along actual flight tracks. *J. Geophys. Res.*, doi 10.1029/D12 2002JD0002515 8524, 2003.

Gauss M, G. Myhre, G. Pitari, M.J. Prather, I.S.A. Isaksen, T.K. Berntsen, Brasseur, G.P. Dentener FJ, Derwent RG, Hauglustaine DA, Horowitz LW, Jacob DJ, Johnson M, Law KS, Mickley LJ, Muller JF, Plantevin PH, Pyle JA, Rogers HL, Stevenson DS, Sundet JK, van Weele M, Wild O, Radiative forcing in the 21st century due to ozone changes in the troposphere and the lower stratosphere. *J. Geophys. Res.*, **108**(D9): Art. No. 4292 May 13, 2003.

Giannakopoulos, C., P. Good, P.K.S. Law, D.F. Shallcross and K.Y. Wang, Modelling the impacts of aircraft traffic on the chemical composition of the upper troposphere, *Proc. Inst. Mech. Eng., J. Aerosp. Eng.*, **217**(G5), 237–243, 2003.

Gheusi, F., J.-P. Cammas, F. Cousin, C. Mari, P. Mascart, Quantification of mesoscale transport across the frontiers of the free troposphere: a new method and applications to ozone. *Atmos. Chem. Phys. Discussion*, Sept. 2004.

Gierens, K.M., U. Schumann, H.G.J. Smit, M. Helten, G. Zängl, Determination of humidity and temperature fluctuations based on MOZAIC data and parametrization of persistent contrail coverage for general circulation models. *Ann. Geophys.* **15**, 1057–1066, 1997.

Gierens, K., U. Schumann, M. Helten, H.G.J. Smit, A. Marenco: A distribution law for relative humidity in the upper troposphere and lower stratosphere derived from three years of MOZAIC measurements. *Ann. Geophys.* **17**, 1218–1226, 1999.

Gierens, K., P. Spichtinger: On the size distribution of ice-supersaturated regions in the upper troposphere and lowermost stratosphere. *Ann. Geophys.* **18**, 499–504, 2000.

Gierens, K., U. Schumann, M. Helten, H. Smit, P.H. Wang, Ice-supersaturated regions and subvisible cirrus in the northern midlatitude upper troposphere. *J. Geophys. Res.* **105**, 22743–22754, 2000.

Gierens, K., R. Kohlhepp, P. Spichtinger, M. Schroedter-Homscheidt, Ice supersaturation as seen from TOVS. *Atmos. Chem. Phys. Discuss.*, **4**, 299–325, (http://www.copernicus.org/EGU/acp/acpd/299/acp-4-299_p.pdf)

Gouget H., G. Vaughan, A. Marenco, H. G. J. Smit., Decay of a cut-off low and contribution to stratosphere-troposphere exchange. *Quart. J. Roy. Met. Soc.*, **126**, 1117–1142, 2000.

Helten M., H., Smit, W. Strater, D., Kley, P., Nedelec, M., Zoger, R. Busen, Calibration and performance of automatic compact instrumentation for the measurement of relative humidity from passenger aircraft, *J. Geophys. Res.*, **103**, 25,643–25,652, 1998.

Helten M., H.G.J. Smit, D. Kley, J. Ovarlez, H. Schlager, R. Baumann, U. Schumann, P. Nedelec, A. Marenco, In-Fight Intercomparison of MOZAIC and POLINAT water vapor measurements, *J. Geophys. Res.*, **104,** D21, 26,087–26,096, 1999.

Law, K.S., P.-H. Plantevin, D.E. Shallcross, H. Rogers, C. Grouhel, V. Thouret, A. Marenco, J.A. Pyle, Evaluation of modelled O_3 using MOZAIC data, *J. Geophys. Res.*, **103**, 25721–25740, 1998.

Law, K. S., P.H. Plantevin, V. Thouret, A. Marenco, W. Asman, W.A.H. Lawrence, P.J. Crutzen, J.F. Muller, D.A. Hauglustaine, M. Kanakidou, Comparison between global chemistry transport model results and measurements of ozone and water vapour by airbus in-service aircraft (MOZAIC) data, *J. Geophys. Res.*, **105**, 1503–1525, 2000.

Lee, S.H., M. Le Disloquer, R. Singh, S.E. Hobbs, C. Giannakopoulos, P.H. Plantevin, K.S. Law, J.A. Pyle and M.J. Rycroft, Implication of NO_y emissions from subsonic aircraft at cruise altitude, *Proc. Inst. Mech. Eng. J. Aerosp. Eng.*, **211**, G3, 157–168, 1997.

Li, Q., D. J. Jacob, J. A. Logan, I. Bey, R. M. Yantosca, H. Liu, R. V. Martin, A. M. Fiore, B. D. Field, B. N. Duncan, V. Thouret, A tropospheric ozone maximum over the Middle East, *Geophys. Res. Lett.*, **28**, 3235–3238, 2001.

Lindborg E., Can the atmospheric kinetic energy spectrum be explained by two-dimensional turbulence?, *J. Fluid. Mech.*, **388**, 259–288, 1999.

Lindborg, E., J. Y.N. Cho, Determining the cascade of passive scalar variance in the lower stratosphere, *Phys. Rev. Lett.*, **85**, 5663–5666, 2000.

Lindborg E. and Cho J. Y.N., Horizontal velocity structure functions in the upper troposphere and lower stratosphere: 2. Theoretical considerations, *J. Geophys. Res.*, **106** (D10), 10233–10241, 2001.

Marenco, A., V. Thouret, P. Nedelec, H. Smit, M. Helten, D. Kley, F. Karcher, P. Simon, K. Law, J. Pyle, G. Poschmann, R. Von Wrede, C. Hume, T. Cook, Measurement of ozone and water vapour by Airbus in-service aircraft: The MOZAIC airborne program, An overview, *J. Geophys. Res.*, **103**, 25631–25642, 1998.

Martin, R.V., D.J. Jacob, J.A. Logan, I. Bey, R.M. Yantosca, A.C. Staudt, Q. Li, A.M. Fiore, B.N. Duncan, H. Liu, P. Ginoux, V. Thouret, Global model analysis of TOMS and in-situ observations of tropical tropospheric ozone, *J. Geophys. Res.*, **107**(D18), 4351, doi: 10.1029/2001JD001480, 2002.

Morgenstern, O., A. Marenco, Wintertime climatology of MOZAIC ozone based on the potential vorticity and ozone analogy. *J. Geophys. Res.*, **105**(D12), 15481–15493, 2000.

Morgenstern, O., G.D. Carver, Comparison of cross-tropopause transport and ozone in the upper troposphere and lower stratosphere region, *J Geophys Res-Atmos* **106**(D10): 10205–10221 May 27, 2001.

Nedelec P., J.-P. Cammas, V. Thouret, G. Athier, J.-M. Cousin, C. Legrand, C. Abonnel, F. Lecoeur, G. Cayez, C. Marizy, An improved infra-red carbon monoxide analyser for routine measurements aboard commercial Airbus aircraft: Technical validation and first scientific results of the MOZAIC III Program. Submitted to *Atmos. Chem. And Phys.*, **3**, 1551–1564, 29-9-2003.

Nedelec P., V. Thouret, J. Brioude, B. Sauvage, J.-P. Cammas, A. Stohl, Extreme CO concentrations in the upper troposphere over North-East Asia in June 2003 from the in-situ MOZAIC aircraft data. *Geophys. R. Letters*, accepted, 2005.

Nedoluha, G.E., R.M. Bevilacque, K.W. Hoppel, J.D. Lumpe, H.G.J. Smit, POAM III measurements of water vapor in the upper troposphere and lower most stratosphere, *J. Geophys. Res.*, **107**, 10.1029/2001JD000793, 2002.

Newell R., V. Thouret, J. Cho, P. Stoller, A. Marenco, H. Smit, Ubiquity of quasi-horizontal layers in the troposphere, *Nature*, **398**, 316–319, 1999.

O'Connor, F.M., K. S. Law, J. A. Pyle, et al., Tropospheric ozone budget: regional and global calculations, *Atmos. Chem. Phys. Discussions*, MS-NR: acp2004–001, 2004.

Offermann D., B. Schaeler, M. Riese, M. Langfermann, M. Jarisch, G. Eidmann, C. Schiller, H.G.J. Smit, W.G. Read, Water vapor at the tropopause during the CRISTA 2 mission, *J. Geophys. Res.*, **107**, DOI 10, 1029/2001JD000700, 2002.

Prados A.I., G.E. Nedoluha, R.M. Bevilacqua, D.R. Allen, K.W. Hoppel, A. Marenco, POAM III ozone in the upper troposphere and lowermost stratosphere: Seasonal variability and comparisons to aircraft observations. *J. Geophys. Res.*, **108**, (*D*7), 10.1029/2002JD002819, 2003.

Prather, M., M. Gauss, T. Berntsen, I. Isaksen, J. Sundet, I. Bey, G. Brasseur, F. Dentener, R. Derwent, D. Stevenson, L. Grenfell, D. Hauglustaine, L. Horowitz, D. Jacob, L. Mickley, M. Lawrence, R. von Kuhlmann, J.-F. Muller, G. Pitari, H. Rogers, M. Johnson, J. Pyle, K. Law, M. van Weele, O. Wild, Fresh air in the 21st century? *Geophys. Res. Letts.*, **30**(2), 1100, doi:10.1029/2002GL016285, 2003.

Ravetta F., G. Ancellet. Identification of dynamical processes at the tropopause during the decay of a cut-off low using high resolution airborne lidar ozone measurements. *Mon. Weather Rev.*, **128**, 3252–3267, 2000.

Roelofs G.J., A.S. Kentarchos, T. Trickl, A. Stohl, W.J. Collins, R.A. Crowther, D. Hauglustaine, A. Klonecki, K.S. Law, M.G. Lawrence, R. von Kuhlmann, M. van Weele, Intercomparison of tropospheric ozone models: Ozone transport in a complex tropopause folding event, **108**(D12): art. no.8529, *J. Geophys. Res.*, 2003.

Sauvage et al., Tropospheric ozone over Equatorial Africa: regional aspects from the MOZAIC data. *Atmos. Chem. Phys.*, **4**, 3285–3322, 2005.

Scott, R.K., J.-P., Cammas, P. Mascart et al., Stratospheric filamentation into the upper tropical troposphere. *J. Geophys. Res.*, **106**(D11), 11835–11848, 2001.

Scott, R.K., J.-P. Cammas, Wave breaking and mixing at the subtropical tropopause. *J. of Atmos. Sciences*. **59**(15), 2347–2361, 2002.

Scott, R.K., E. Schuckburgh, J.-P. Cammas, B. Legras, Stretching rates and equivalent length near the tropopause. *J. of Geophys. Res.*, D13 2002JD002988, 4394.

Spichtinger P., K. Gierens, W. Read, The statistical distribution law of relative humidity in the global tropopause region. *Meteorol Z*, **11**(2), 83–88, 2002.

Spichtinger, P., K. Gierens, U. Leiterer, H. Dier, Ice supersaturated region over Lindenberg, Germany. *Meteorol. Z.*, **12**, 143–156, 2003.

Spichtinger, P., K. Gierens, W. Read, The global distribution of ice-supersaturated regions as seen by the Microwave Limb Sounder. *Q. J. R. Meteorol. Soc.*, **129**, 3391–3410, 2003.

Stohl, A., T. Trickl, A textbook example of long-range transport: Simultaneous observation of ozone maxima of stratospheric and North American origin in the free troposphere over Europe. *J. Geophys. Res.*, **104**, 30445–30462, 1999.

Stohl, A., P. James, C. Forster, N. Spichtinger, A. Marenco, V. Thouret, H. G. J. Smit, An extension of measurements of ozone and water vapor by Airbus in-service aircraft (MOZAIC) ozone climatologies using trajectory statistics. *J. Geophys. Res.* **106**, 27,757–27,768, 2001.

Suhre K., J.-P. Cammas, P. Nedelec, R. Rosset, A. Marenco, H.G.J. Smit, Observations of high ozone transients in the upper Equatorial Atlantic troposphere. *Nature*, **388**, 661–663, Aug. 1997.

Spichtinger, P., K. Gierens, H.G.J. Smit, J. Ovarlez, J.-F. Gayet, On the distribution of relative humidity in cirrus clouds. *Atmos. Chem. Phys. Discuss.*, **4**, 365–397, 2004. (http://www.copernicus.org/EGU/acp/acpd/365/acpd-4-365_p.pdf).

Thouret, V., A. Marenco, J. A. Logan, P. Nedelec, C. Grouhel, Comparisons of ozone measurements from the MOZAIC airborne program and the ozone sounding network at eight locations, *J. Geophys. Res.*, **103**, 25,695–25,720, 1998.

Thouret, V., A. Marenco, P. Nédélec, C. Grouhel, Ozone climatologies at 9–12 km altitude as seen by the MOZAIC airborne program between September 1994 and August 1996, *J. Geophys. Res.*, **103**, 25,653–25,679, 1998.

Thouret V.J.Y.N. Cho, R.E. Newell, et al., General characteristics of tropospheric trace constituent layers observed in the MOZAIC program. *J. Geophys. Res.*, **105**(D13), 17379–17392, 2000.

Thouret V, J.Y.N. Cho, M.J. Evans, et al., Tropospheric ozone layers observed during PEM-Tropics B. *J. Geophys. Res.*, **106**(D23), 32527–32538, 2001.

Thouret et al., Tropopause-referenced ozone climatology and interannual variability (1994–2000) from the MOZAIC program. Submitted to *Atmos. Chem. Phys. Discuss*, 2005.

Tulet P, K. Suhre, C. Mari, et al., Mixing of boundary layer and upper tropospheric ozone during a deep convective event over Western Europe. *Atmos. Environ.*, **36**(28), 4491–4501, 2002.

Zahn, A., C.A.M. Brenninkmeijer, W.A.H. Asman, P.J. Crutzen, G. Heinrich, H. Fischer, J.W.M. Cuijpers, P.F.J. van Veldhoven, The budget of O_3 and CO in the upper troposphere: The CARIBIC aircraft results 1997–2001. *J. Geophys. Res.* **107**(D17), art. no. 4337, 2003.

Zbinden R., J.-P. Cammas, V. Thouet, P. Nédélec, F. Karcher, P. Simon, Midlatitude tropospheric ozone columns from the MOZAIC programme: climatology and interannual variability. Submitted to *Atmos. Chem. And Physics Discussion*, 2005.

Chapter 8

Uninhabited Aerial Vehicles: Current and Future Use

PAUL A. NEWMAN

Introduction

Uninhabited aerial vehicles (UAVs) have recently become available for scientific research. UAV development has largely been underwritten by the military for reconnaissance. UAVs such as the Global Hawk have demonstrated a high-altitude (>16 km), long-range (>2000 km), and heavy payload (<1000 kg) capability. These attributes make UAVs very attractive for Earth science research.

A UAV is classified as a powered vehicle that does not carry a human operator, uses aerodynamic forces to provide vehicle lift, can fly autonomously or be piloted remotely, can be expendable or recoverable, and carries a payload. Remotely piloted vehicles (RPVs) are a subclass of UAVs that require a ground pilot to control the aircraft. Radio-controlled aircraft are not UAVs if they do not carry a payload. Satellites are not UAVs, since they do not use aerodynamic forces. On the other hand, satellites bear many similarities to UAVs, as will be discussed later.

In this paper, we will first discuss the usage of conventional aircraft for scientific missions, and how conventional aircraft relate to both satellites and UAVs. In the third section we will discuss some specific examples of UAVs, give a recent history of UAV usages for Earth science and applications, and give a brief description of UAV operations. The fourth section will discuss advantages and disadvantages of UAVs, and the fifth section will discuss future directions for UAV usage.

Aircraft

How Are Aircraft Used for Earth Observations?

Earth scientists have extensively used aircraft, and their use continues to grow. Currently, aircraft are used for both remote sensing and in-situ sampling of the atmosphere. Aircraft are used for (1) field experiments such as the Cirrus Regional

Study of Tropical Anvils and Cirrus Layers—Florida Area Cirrus Experiment (CRYSTAL-FACE), (2) monitoring and applications (e.g., forest fire overflights and reconnaissance), (3) satellite validation (e.g., SOLVE II), (4) instrument and algorithm development, and in the future, (5) as suborbital systems that are integrated into satellite programs.

Aircraft are used to measure long-term events that have time scales of months to years, short-term events with time scales of days to weeks, and mesoscale events with time scales of minutes to hours. Examples of long-term events are the Antarctic ozone hole, pollution systems, and climate change. These events are continuous or typically recur on a regular basis. Long-term events have large spatial scales and are easily planned for. Short-term events are typically synoptic scale (1000–5000 km). Streamers of pollution or synoptic scale weather systems are classic examples. While these short-term events are not precisely predictable, they tend to regularly recur, and can be planned for. They are typically more difficult to sample than the long-term, large-scale events. The mesoscale systems are the most difficult to sample. A convective system with a spatial scale of 10 km and a time scale of 3–4 hours is extremely difficult to sample. The transit time from the airfield to the phenomena may see the entire development and collapse of these systems. The smallerscale phenomena with chaotic behavior are the most difficult phenomena to sample. In any case, all of these phenomena require clever planning and good forecast tools.

The CRYSTAL-FACE (CF) mission provides an interesting example of how aircraft are used to sample phenomena. CF was a measurement campaign for investigating tropical cirrus cloud properties and formation. CF was conducted in July 2002 from Key West, Florida, and included a mix of aircraft, ground-based observations, radiosondes, and satellite instruments. CF used 6 aircraft: the NASA ER-2, the NASA WB-57F, the Scaled Composites Proteus, the University of North Dakota's Citation, the CIRPAS Twin Otter, and the Naval Research Lab's P-3 Orion. The CF location was chosen based upon the regular early afternoon daily convection that occurs in South Florida over the moist Everglades. This convection was driven by the hot and moist boundary that was triggered by the daily sea breeze. This convection detrained cirrus at about 13 km. A surface radar in the Everglades tracked the development of this convection and directed the aircraft to sample the evolving cirrus as it detrained.

The ER-2 payload (Figure 1) was principally composed of remote sensing instruments, while the WB-57F (Figure 2) payload was composed of in-situ instruments. The ER-2 flew above the detraining cirrus at an altitude of 20 km, while the WB-57F typically flew in the cirrus. The ER-2 and WB-57F were flown together along the same track (ER-2 above and WB-57F below), such that they were sampling the same cirrus simultaneously.

The ER-2 and WB-57F were also flown to validate the NASA Aqua, Terra, and TRMM satellite instruments. Since the ER-2 duplicated the capabilities of these satellites (see satellites in Figure 1), it was possible to both validate satellite remote sensing instruments directly, and to bootstrap the in-situ observations to the satellite observations using a forward radiative transfer model. This was

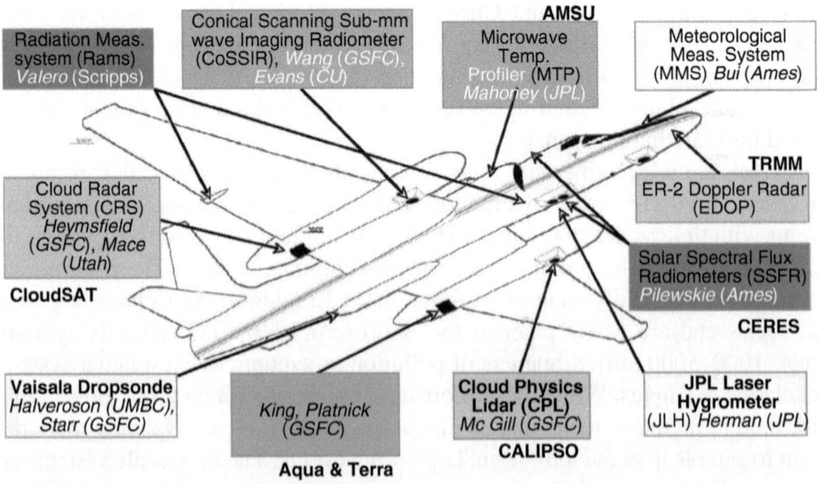

FIGURE 1. NASA ER-2 payload diagram for the July 2002 CRYSTAL-FACE mission. Total payload was 2183 lbs (992 kg).

accomplished by directly comparing the in-situ observations (e.g., ice water content) to the ER-2 measurements, and then to the same measurement made by the satellite.

The CF mission demonstrated key aspects of aircraft field missions. First, it was a clever choice of location for the investigation of cirrus detrainment. Second, it

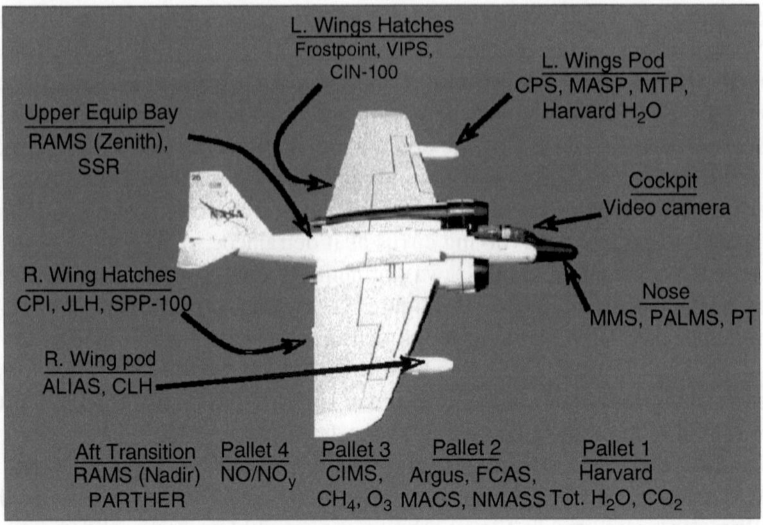

FIGURE 2. NASA WB-57F payload diagram for the July 2002 CRYSTAL-FACE mission. Total payload was approximately 4000 lbs (1820 kg).

showed the need for excellent forecasting and nowcasting for directing flight operations. Third, they optimally used the platforms such that each platform's data could be related to the other platforms. Fourth, they combined ground, aircraft, and satellite observations into a unified experiment. Finally, CF clearly demonstrated how aircraft could be operationally integrated with satellite systems.

The challenge for UAVs is to both duplicate some of our current capabilities with manned aircraft, and to extend these capabilities.

How Do Aircraft Relate to Other Satellites?

Satellites and aircraft have different spatial coverage, temporal coverage, resolution, lifetimes, and payloads. Earth sciences satellites are typically in low Earth orbits (LEO) or geosynchronous Earth orbits (GEO). The LEO orbits of satellites such as Aura or Terra give approximately twice per day global coverage using scanning type instruments with day/night viewing capability. For a specific target such as a cloud or ground site, the LEO satellite provides 2 views per day. A GEO orbiting satellites such as GOES-8 provides continuous coverage for about 1/6th of the Earth's surface. In contrast, an aircraft can provide approximately 6–8 hours of continuous observations of a specific location. LEO orbits are relatively low (e.g., Aura orbits at an altitude of about 780 km); GEO orbits are very high (36,000 km). For trace gas observations, aircraft can provide in-situ observations with horizontal/vertical resolutions of 0.1/0.01 km; GEO instruments generally provide horizontal/vertical resolutions of 1/5 km, while LEO instruments have resolutions of 10/1 km. Aircraft make sporadic observations depending on campaigns, while satellites provide near-continuous observations over 2–10-year periods. Satellite payloads are typically a few kg to 2000 kg, and aircraft payloads are typically a few hundred kg to 15,000 kg for aircraft such as the NASA DC-8.

Satellites and aircraft provide somewhat different capabilities. However, as was shown in the CF mission example, these platforms are mutually dependent. Because of their range and duration, UAVs begin to approach the capabilities of satellites but with a capability of providing much more detailed observations of specific phenomena.

How Do Manned Aircraft Relate to UAVs?

Manned aircraft and uninhabited aerial vehicles (UAVs) have many similarities, but also have significant differences. The advantages of manned aircraft over a UAV are their (1) long heritages, (2) wide varieties, (3) immense payload, power, and volume capacities, (4) airspace access, (5) real-time command and control, and (6) ability to carry passengers and instrument operators. The drawbacks for manned aircraft are (1) short duration capabilities (≤ 10 hours), (2) modest range (< 3700 km), (3) inability to fly in dangerous situations, and (4) inability to fly in dirty environments (ash, nuclear debris, etc.).

UAVs provide opportunities to overcome some of the shortcomings of manned aircraft. First, some UAVs are capable of very long-duration and long-range flights (> 5,000 km, > 24 hours). Second, some UAVs are expendable. Third, UAVs can be flown in dirty conditions where a pilot could be harmed. UAVs are capable of doing dumb, dangerous, and dirty missions. In particular, because of the shorter durations for manned aircraft, manned aircraft are best suited for reconnaissance (i.e., short snap-shot views of phenomena). Because of its longer endurance, the UAVs expand our capabilities to perform surveillance (i.e., continuous observations of phenomena). Figure 3 is a plot of aircraft range and altitude for a number of UAVs and manned aircraft. Also included in Figure 3 are the horizontal and vertical dimensions of various atmospheric phenomena. Some UAVs (e.g., the Global Hawk) significantly expand the envelope beyond our current manned aircraft capability.

Uninhabited Aerial Vehicles (UAVs)

Examples

There are dozens of UAVs that have been developed or are under development today (see: http://www.uavforum.com/vehicles/capabilities.htm or http://www.aiaa.org/images/PDF/WilsonChart.pdf). These UAVs have been mainly developed for military purposes. Two broad classes of UAVs are high-altitude, long-endurance

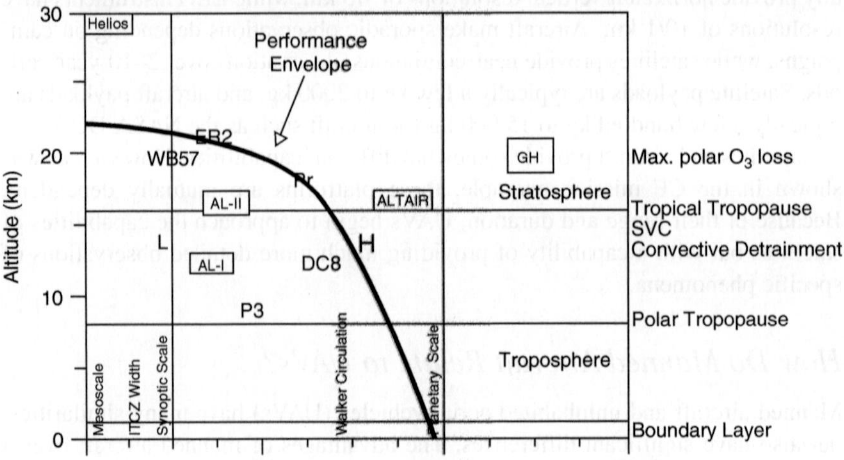

FIGURE 3. Altitude versus range plot for manned aircraft and UAVs (in boxes). The aircraft payload weight is proportional to font size (truncated at 2000 lb and 600 lb), and the bold-faced font indicates payloads greater than 2000 lb. The thick line indicates the performance envelope for manned aircraft. The L indicates a Learjet, while the H indicates the NSF Hiaper Gulfstream V aircraft. The scales of various atmospheric phenomena are indicated for perspective.

(HALE) UAVs and medium-altitude, long-endurance (MALE) UAVs. In general, a long-endurance is defined to be greater than about 10 hours, while a high-altitude is defined to be about about 50,000 feet (15.2 km). A number of UAVs are now used or have been used for Earth Sciences and applications. Table 1 shows a brief listing of UAVs that have been or have planned uses for Earth science research.

The HALE Global Hawk has some of the most desirable characteristics: a long endurance, high altitude, and long-range capability. The Global Hawk is powered by a conventional turbofan jet engine. In contrast, the Helios and the Pathfinder Plus (Figure 5) are solar-powered with large wings covered in solar cells. Because they carry solar cells and some batteries, they have tremendous altitude capability but relatively short range (daytime flights only) and small payloads. The Aerosonde falls into the "small" UAV class. It has a wingspan of only 2.9 meters and is powered by at 24cc engine that runs on unleaded gasoline fuel. The Aerosonde is launched from the rooftop of a speeding car, while the other UAVs require conventional runways.

History and UAV Limits

The characteristics of UAVs are engineered to suit specific roles. UAVs originally evolved from target drones. The U.S. military extensively used the first UAVs during the 1960s in Vietnam [Wilson, 2003]. The Firebee and Compass Arrow programs developed these UAVs for reconnaissance, but the programs largely ended after the war. During the 1991 Gulf War, UAVs began to show their potential for flying in hostile conditions. UAVs are now extensively used by militaries around the world, and are being further developed by many countries.

As with manned aircraft, there is a tradeoff between maximum altitude, payload, range, and airspeed. The lift provided by a wing is proportional to the area of the wing, the air density, and the square of the aircraft's airspeed. Typically we measure the lift capability of an aircraft by the "wing loading." Wing loading is the ratio of the weight of an aircraft to the area of the wings. A UAV with a low wing loading can fly at higher altitudes (lower density) with airspeed less than the speed of sound. A high altitude aircraft such as the Helios had a large wing and a low weight, giving it very small wing loading of 0.81 lbs./ft^2. This low wing

TABLE 1. Selective UAVs That Have Been Used or Proposed for Science Missions.

UAV (manufacturer)	Range (km)	Endurance (hrs)	Ceiling (km)	Payload (kg)
Global Hawk (Northrop Grumman)	25,000	36	19.8	910
Helios (Aeroenvironment)	200	15	30.0	16
Pathfinder (Aeroenvironment)	200	14	24.5	11
*Proteus (Scaled Composites)	5000	14	16.7	1000
Altair (General Atomics)	4200	32	15.2	300
Altus-II (General Atomics)	5600	24	13.7	150
Aerosonde (Aerosonde)	3000	30	4.0	5

Numbers are taken or calculated from manufacturers specifications. Actual performance will differ from these estimates. *Proteus is a manned aircraft that has only been used in a UAV demonstration.

loading allowed the Helios to achieve altitudes near 100,000 feet. As payload weight is added, the wing loading increases and the aircraft can't fly as high. In contrast to the Helios, the Global Hawk has a gross weight of 25,600 lbs and a wing area of 540 ft^2, giving a wing loading of 47 lbs./ft^2. The Global Hawk flies much faster than the Helios to gain the lift necessary to fly at 60,000 feet. The Global Hawk has an airspeed of about 343 kts at 20 km, while the Helios has an airspeed of about 150 kts at 30 km. Adding payload reduces the maximum altitude attainable by these aircraft.

The drawback of a low wing loading comes from the necessity for a large wing. A UAV with large wing and a low weight is very susceptible to ground weather and cross winds. A takeoff becomes extremely difficult, since the aircraft wing acts as a large sail that can easily overturn the very light UAV. This creates the necessity for a very calm wind for both takeoff and landing. Airfields with extremely calm winds over significant periods are difficult to find.

Operations

UAV operations are similar to operations for manned missions. However, UAVs require different command and control.

The General Atomics MALE Altus aircraft (Figure 4) provides a useful example of operational constraints for UAVs. The Altus is a modestly size UAV with a length of about 7 m and a wingspan of 17 m. The Altus is powered by a gasoline engine, and can carry a payload of 150 kg to an altitude of about 14 km and has a range of about 5600 km. It has a C-band line-of-sight data link with a GPS/INS system. The Altus is easily disassembled into shipping cases, so it is easily transported and reassembled. The Altus is an excellent example of a versatile aircraft for atmospheric sampling.

The Altus can be remotely piloted, or it can be run just like a satellite without human intervention. The waypoints are input to the program, and the aircraft executes those commands. The UAV can be communicated with using Ku-band

FIGURE 4. General Atomics Altus-I in flight while retracting landing gear in August 1997 (NASA DFRC, photographer Carla Thomas).

(satellite) and C-band (microwave) communications to both direct and control the flight. While the Predator has a Ku-band link provides over the horizon communications, the Altus is not currently configured for over-the-horizon communications. The Altus has a forward-looking camera, such that flying the UAV is like sitting in an aircraft cockpit, but looking forward with narrow peripheral vision. Trailers containing the ground communications system with seating for scientists and the ground pilot are set up at the UAVs location, although the UAV can also be controlled from far distant locations because of the satellite link.

UAVs as Science Platforms

UAVs in Field Experiments

A few field campaigns have now been conducted with UAVs. In particular, the use of UAVs for Earth sciences and applications has been led by three programs: The Department of Energy's Atmospheric Radiation Measurement (DOE/ARM) Program, the NASA Environmental Research Aircraft and Sensor Technology (ERAST) program, and the NASA UAV Science Demonstration Program (UAVSDP), and NASA's Earth Sciences directorate.

The DOE/ARM program has flown a number of UAV missions [Stephens et al., 2000]. Since 1993, DOE/ARM has sponsored 4 flight campaigns using 3 General Atomics UAVs: the Gnat-750, Altus-I, and Altus-II. The 1st science flight of a UAV was accomplished by the Gnat-750 in April 1994 when it made radiative flux measurements to an altitude of 7 km. They further flew a 26-hour duration flight over the CART site (5 Oct. 1996). In addition to these flights, the DOE/ARM program has developed several compact instruments suitable for UAV applications.

The NASA ERAST program was started in 1994 to develop UAV technology and instrumentation. This program led to the development of a number of small instruments, and participated in the development of several high altitude aircraft. ERAST facilitated the development of (1) the Aurora Flight Sciences' Perseus aircraft, which achieved a flight altitude of 60,260 feet in June 1998, (2) flights of the General Atomics Aeronautical Systems' Altus II to 55,000 ft., with joint DOE/ARM radiation experiments at Kauai, Hawaii, and wildfire imaging demonstration in September, 2001, and (3) flights of the solar-powered AeroVironment's Pathfinder-Plus (Figure 5) and HELIOS, with a flight of the HELIOS to 96,863 feet from Kauai in August 2001.

The NASA UAVSDP funded two science missions in 2001: the Altus Cumulus Electrification Study (ACES), and the UAV Coffee Project. The ACES objectives were to (1) validate a satellite instrument (the Lightning Imaging Sensor), (2) investigate lightning type, cloud-top optical energy, and power statistics, (3) lightning-storm relationships, and (4) storm electric budgets. The mission was flown in August 2002 from Key West, Florida. The Altus-II completed 13 flights and acquired 30 hours of data (http://aces.msfc.nasa.gov/index.html). The UAV

FIGURE 5. AeroVironment's Pathfinder solar-powered plane in flight on September 11, 1995 (NASA DFRC, photographer: Tony Landis).

coffee project was flown in September 2002 from Kauai using AeroVironment's Pathfinder-Plus [Herwitz et al., 2004]. The project's objective was to image the Kauai Coffee Company's coffee farms for the detection of differences in coffee field ripening. The Pathfinder conducted a single 12-hour flight on September 30, 2002. The Pathfinder payload was an RGB digital camera and a narrow band digital multispectral camera.

The Aerosonde has been used in a number of experiments (http://www.aerosonde.com/operations/). Two examples are the Fourth Convection And Moisture EXperiment (CAMEX-4), which was flown in the fall of 2001 and was sponsored by the NASA Earth Science Enterprise.. CAMEX-4 included 8 flights of the Aerosonde for studying the boundary at very low altitudes. The Aerosonde has also been flown from Barrow, Alaska, to study sea surface temperatures in the Arctic (Inoue and Curry, 2004).

Use of UAVs for science in Europe has also been demonstrated with some flights of the EADS/IAI Eagle UAV from Kiruna, Sweden. The Eagle flights objectives were (1) give the Swedish military operational experience, (2) allow the Swedish Space Corporation to study UAVs as scientific platforms, (3) provide the Meteorological Institute (Univ. of Stockholm) UAVs for atmospheric sampling experience (Condensation Particle Counter). The flights of the Eagle were performed in June 2002 with 6 flights totaling 25 hours, that included one science flight of 5 hours (Hedlin and Abrahamsson, 2003).

UAV Problems and Current Uses

UAVs present rather unique problems. First, at present they are unreliable. The General Atomics' Predator RQ-1A had a mishap rate of 43 per 100,000 hours, and the RQ-1B had 31 [Peck, 2003]. This suggests that an Altus would be lost approximately once in every 3,000 flight hours. Such loss rates are 10 times larger than loss rates for current high-performance manned fighter aircraft. The NASA ERAST program was evaluating the Aurora Perseus B aircraft when it crashed on Oct. 1, 1999 (http://www.space.com/news/perseus crash991004.html). The one-of-a-kind AeroVironment HELIOS crashed on June 26, 2003, and is not currently scheduled to be replaced. Such mishaps demonstrate that UAVs are still under development. While some UAVs are expendable, they are not randomly expendable. While inexpensive instruments on hazardous flights might be risked, test flights of one-of-a-kind multi-million dollar instruments cannot be risked.

Second, aircraft must be flown frequently. The typical ER-2 or WB-57F can generally fly every day, but more typically fly every other day. For a typical 14-day deployment, 5–12 flights totaling 40–100 hours can be flown. UAVs require more maintenance and can be more susceptible to weather conditions. During ACEs, the Altus-II was operational from August 2–30 (29 days), but only achieved a total of 30 hours of data. A substantial amount of these days were lost because of a problem with the specialized engines used by the Altus-II. The solar planes have extremely large wings, making them very susceptible to crosswind conditions on takeoff. During the UAV coffee project, AeroVironment's Pathfinder-Plus had to scrub numerous times because of weather (http://geo.arc.nasa.gov/uav-nra/index.html).

Third, UAVs are not low cost. A UAV eliminates the necessity for a pilot and all of his life support, creating a weight dividend that can be used for fuel or payload. However, a UAV still requires complex communication with a ground system. This means that there is no "personnel" dividend. A UAV still requires ground support with a ground-pilot, mechanics, electricians, and computer specialists. Because UAVs are development aircraft, they typically have redundant command-and-control systems. This reduces the weight dividend that was gained by the absence of the pilot. Further, because UAVs are development planes, they have higher staffing than manned aircraft with established heritages. The cost of flying UAVs is higher than a manned aircraft because of (1) more personnel, (2) more equipment to operate the UAV, (3) UAVs are still under development, and (4) high insurance rates.

Fourth, UAVs currently do not carry sufficient payload in comparison to manned aircraft. Much of the benefit of a manned aircraft is its ability to carry a number of heavy instruments. During the SOLVE-I campaign, the ER-2 carried 21 instruments, making it a functional chemical kinetics laboratory. The NASA WB-57F can carry 2300 kg to 19 km, and the NASA DC-8 can carry 13,600 kg to 12.5 km. As is shown in Table 1, the Global Hawk can carry a 910 kg payload to 19.8 km. The Global Hawk has the largest payload capacity of UAVs, but is considerably short of manned aircraft capacity. Finally, antennas and other equipment for redundant and robust aircraft control (e.g., over-the-horizon capabilities) consume valuable payload weight and volume.

Fifth, UAVs are both very lightweight and typically built of composite materials. This creates problems for flying in severe weather conditions. For example, the crash of the Helios was attributed to turbulence in the lee of the island of

Kauai. The turbulence caused the aircraft's wing to abnormally bend beyond their breaking point. UAVs also cannot be flown where lightning strikes are probable, since lightning will melt or burn the composite material.

Finally, UAVs have great difficulty gaining access to airspace. This is a severe restriction for Earth science. UAV flight request requires an extended period before the FAA approval. "Certified UAVs," such as the Global Hawk, require only a few days, but are limited to take off and landing in restricted airspace. Nevertheless, atmospheric phenomena can move thousands of kilometers over the few days between the flight request and its approval. Good forecasting improves this, but errors of hundreds of kilometers would be expected.

Many of these UAV problems will eventually be solved. However, it is clear that UAVs cannot currently replace manned aircraft. Developmental instruments that require a "hands-on" human operator are flown on manned aircraft such as the DC-8. This will probably never be done using a UAV. The UAVs currently being flown do not have the lift capability of aircraft such as the DC-8, WB-57F, or ER-2. There are no heavy lift UAVs in development.

In spite of problems, NASA and other agencies continue to promote UAVs for Earth sciences. NASA and NOAA flew the General Atomics MALE Altair in the spring of 2005 for Earth science research probably out over the Pacific Ocean. NASA's Research, Education and Applications Solutions Network (REASoN) has jointly developed a program with the U. S. Department of Agriculture to also use the Altair for monitoring forest fires in the Western United States.

The spring 2005 NOAA and NASA mission flew the Altair from NASA Dryden Flight Research Center. The purpose of this mission was to test instruments and NOAA operational requirements. The Altair payload included remote and in-situ instruments for measurements of ocean color, and atmospheric composition and temperature; and a surface imaging system. In-situ composition measurements included ozone and long-lived gases such as halocarbons and nitrous oxide. The vertical distribution of water vapor has been remotely measured with passive microwave sensors. Six flights have been performed for a total of 53 hours of flight time. Flights reached altitudes up to 45,000 ft and have durations up to 20 hours. Flight objectives included sampling low-level jets in the eastern Pacific Ocean that bring moisture to the continental US; sampling regions of high potential vorticity at mid-latitudes that result from transport of polar air; and imaging of the Channel Islands National Marine Sanctuary (CINMS) to examine shorelines and evaluate the potential for marine enforcement surveillance. This NOAA/NASA experiment was a prelude to future routine operations that are targeted on improving both Earth science research, operational forecasting of both weather and pollution, and applications such as disaster assessment and mitigation.

Future Directions

The future of UAVs in Earth science is bright, but the use of UAVs for Earth science is only in its infancy. The long-range of UAVs such as the Global Hawk and Altair bring phenomena within our reach that were unachievable in past decades. It is becoming possible to monitor the vast reaches of the Central Pacific or view Arctic Sea ice changes with a single aircraft launched from Hawaii or California. Slowly moving hurricanes, typhoons, and cyclones could be monitored and fully surveyed for tens of hours (surveillance), rather than with limited over flights by manned aircraft (reconnaissance). The extended duration of UAVs provides a new surveillance capability for Earth scientists.

Most shortcomings of UAVs are being solved. Safety procedures, improved components and software, and continued military developments will increase the reliability of UAVs. New initiatives such as "Access 5" will address FAA flight restrictions and provide access to U.S. national airspace for UAVs. Routine access by Earth scientists to UAVs and access by UAVs to airspace will become common in the next few decades.

UAVs bridge the temporal gap between the twice per day LEO satellite observations. UAVs also can provide the high vertical resolution and polar coverage that can't be obtained by GEO satellite observations. The shorter endurance of manned aircraft makes them capable of performing reconnaissance (i.e., high-resolution snapshots of phenomena) while the long endurance of UAVs makes them capable of remaining with phenomena and providing surveillance (i.e., high-resolution continuous observations of a phenomena). In addition to their capability to fill a gap in the satellite coverage, UAVs provide the capability to enhance satellite remote sensing with high-resolution in-situ observations. These in-situ observations allow the satellite "radiance" to be built upward from forward model of radiative transfer. UAVs will evolve into "sub-orbital" satellite systems.

In summary, manned aircraft and UAVs provide high-resolution observations that are (1) synergistic with satellite observations, and (2) fill observation gaps in satellite coverage. UAVs show the promise for major improvements in range and duration for Earth science. This improved capability moves UAVs from a reconnaissance role to a surveillance role. The improved capability also provides for a greatly enhanced capability of targeting phenomena. Scientists are using UAVs, and field missions have demonstrated the utility of UAVs. In the future, UAV missions will extend the scope of traditional manned-aircraft missions to those that are ultra-long, in dirty environments, and dangerous environments. In particular, these long-range, extended-duration UAVs will provide a logical extension of satellite missions to autonomous environmental monitoring as a part of sensor webs.

References

Stephens, G.L., et al., The Department of Energy's Atmospheric Radiation Measurement (ARM) Unmanned Aerospace Vehicle (UAV) program, *Bull. Amer. Met. Soc.*, **81**, 2915–2937, 2000.

Hedlin, S., M. Abrahamsson, Demonstration of Eagle MALE UAV for scientific research, Proc. 18th Bristol International Conference on Unmanned Air Vehicle Systems, Dept. Aerospace Eng., University of Bristol, Bristol, United Kingdom, 31 March–2 April 2003.

Herwitz, S.R., et al., Imaging from an unmanned aerial vehicle: Agricultural surveillance and decision support, *Comp. and Electr in Agri.*, **44**, 49–61, 2004.

Inoue, J., A. Curry, Application of Aerosondes to high-resolution observations of sea surface temperature over Barrow Canyon, *Geophys. Res. Lett.*, **31**, DOI: 10.1029/2004GL020336, 2004.
Peck, M., Pentagon Unhappy About Drone Aircraft Reliability, National Defense, 2003.
Wilson, J.R., UAVS: A WORLDWIDE ROUNDUP, Aerospace America, 2003.

Chaper 9

U.S. Ground-based Campaign—PM Supersite Program

KENNETH L. DEMERJIAN

Introduction

The periodic assessment of new scientific information on the health and welfare effects of criteria air pollutants is required under the Clean Air Act and its amendments. As a result of new scientific information identified in its most recent particulate matter (PM) Criteria Assessment Document, EPA revised the PM National Ambient Air Quality Standard (NAAQS) in 1997 (U.S. EPA, 1997) Among the findings presented in this critical assessment, was the following.

> Overall, there is strong epidemiological evidence linking (a) short-term (hours, days) exposures to PM2.5 with cardiovascular and respiratory mortality and morbidity, and (b) long-term (years, decades) PM2.5 exposure with cardiovascular and lung cancer mortality and respiratory morbidity. The associations between PM2.5 and these various health endpoints are positive and often statistically significant. There is also extensive and convincing evidence for associations between short-term exposures to PM10 and both mortality and morbidity.

As a result of the PM2.5 (NAAQS) 1997 revision the U.S. Congress in response to public health concerns and in recognition of the uncertainty associated with some key aspects of the science of PM, decided in 1998 to provide funds for major research initiatives to address several policy-relevant scientific questions associated with PM2.5 regulation. Congress recognized that unlike gaseous pollutants, is not a single compound, but rather a complex mixture whose composition and morphology can vary in time and space and that these airborne particles have many sources and contain hundreds of inorganic and organic compounds. It also recognized that the new PM2.5 standard would placemany metropolitan areas in the U.S. into nonattainment status of the NAAQS. Figure 1 shows annual PM2.5 mass concentrations from 1007 locations in the U.S. for the period 1998–2000 [NARSTO, 2003]. The reds circles indicate locations exceeding the PM2.5 NAAQS. The yellow circles are locations likely to be violation in the near term, considering typical growth patterns. The significance of the magnitude of violations was another reason for Congressional interest.

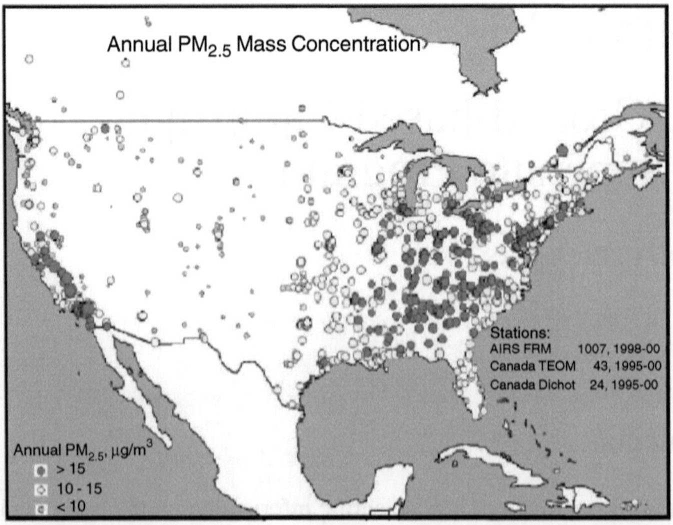

FIGURE 1. Average PM2.5 concentrations in U.S and Canada (source Figure 6.7, NARTSO, 2003).

With these matters in mind, the Congress wanted the community to address two major PM science policy-relevant questions. Which component(s) of the PM mix is/are responsible for the statistical association with adverse health impacts and what are the mechanisms of action? What is the relationship between aerosol and aerosol precursor emissions and exposure of sensitive populations to the components identified above? As part of this research initiative a major workshop was convened involving the scientific and stakeholder communities. Its goal was to identify the key components and design parameters for a comprehensive measurement program to characterize ambient PM and important co-pollutants, in a way that optimizes information for multiple disciplines. A strategic ambient monitoring research program designed to develop, deploy, and evaluate measurement technologies for the monitoring the physical and chemical characteristics of particulate matter (PM) and its relationship to PM mass as measured by the Federal Reference Method (FRM). A competitive request for proposals resulted in the funding of series of programs. These measurement/research programs consisted of two Phase I and seven Phase II so-called PM Supersites, distributed geographically across the country (http://www.epa.gov/ttn/amtic/supersites.html) PM Supersites. Although quite diverse in their approaches, the "PM Supersite" programs shared several common themes across the country. These common themes were (1) to obtain atmospheric measurements to characterize PM, its constituents, precursors, co-pollutants, atmospheric transport, and source categories that effect the PM in the region, (2) obtain atmospheric measurements to support health effects and exposure research and to address the research questions and scientific uncertainties about PM source-receptor-exposure effects relationships, and (3) to conduct methods testing to compare and evaluate new PM measurement technologies for characterizing PM (physically and chemically). Each Supersite had specialized themes as well, including such activities as (1) characterizing PM urban/rural and summer/winter contrasts, (2) characterization PM diesel emission control technologies; PM organics speciation and transformations, (3) sources of new particle formation, (4) development and evaluation of deterministic and source apportionment modeling systems, (5) integrating and linking advanced PM size and chemical composition measurements with on going health studies, and (6) spatial characterization of ultrafine particles in the vicinity of major highways. The PM2.5 Supersite Phase I programs started in 1999, Phase II programs in 2000. The Phase I programs were typically 1–2 years in duration, while the Phase II programs operated over a 3–5 year period. Significant results have already been reported in the peer-reviewed literature (~147 papers) with many more in review and in preparation. Some specific results from the PM2.5 Technology Assessment and Characterization Study in New York (PMTACS-NY), one of the seven Phase II PM Supersite programs are briefly discussed as a representative sample of the measurements/research performed under the program. The PMTACS-NY program to date has provided significant results addressing its three major objectives:

- To measure the temporal and spatial distribution of the PM2.5/co-pollutant complex including photochemical oxidants and precursor species, aerosol size distribution and composition in urban and regional environments.
- To monitor the effectiveness of new heavy-duty vehicle emission control technologies [i.e., Compressed Natural Gas (CNG) and retrofit Diesel Filter—Continuously Regenerating Technology (DF-CRT)] introduced by the NYC Transit Authority for its MTA bus fleet and its potential impact on ambient air quality.
- To test and evaluate new measurement technologies and provide tech-transfer of demonstrated operationally robust technologies for routine network operation.

In the area of instrument evaluation, laboratory and field-based performance evaluations of water management system on continuous PM2.5 mass (R&P SES-TEOM) monitoring systems [Schwab et al., 2003, 2004b, 2004c] have been performed. These studies have demonstrated that marked improvements have been made in reducing volatile losses of PM2.5 as compared to the standard 50 ° TEOM mass monitor. The studies also showed that mass differences are seasonal and likely associated with the content of NH_4NO_3 and semivolatile organic compounds in the PM. Based on these results recommendations regarding operational deploy of SES-TEOM systems and new generation Filter Dynamic Measurement System (FDMS TEOM) technology in routine PM2.5 monitoring networks have been made. In addition to the continuous mass monitors, semi-continuous PM2.5 sulfate and nitrate species measurement systems [Drewnick et al., 2003, Hogrefe et al., 2004, Hering et al., 2004] have been evaluated. These included advanced research measurement systems [Aerosol Mass Spectrometer (AMS), Particle-In-Liquid Sampler (PILS-IC)] as well as commercially viable systems [R&P 8400N/S and Continuous Ambient Sulfate Monitor (CASM)]. The CASM system is currently under development at Thermo, Inc. (under the designation 5020 Sulfate Particulate Analyzer). Over the course of these studies a variety of operational and maintenance issues have been identified and fed back to the instrument manufacturers, resulting in improvements in both the hardware and software of the instrumental systems. Overall the performance evaluations have been quite successful, demonstrating that the routine monitoring systems (R&P 8400S and 8400N; CASM) although having some systematic biases as compared to filter-based measurements can provide good-quality semi-continuous PM2.5 sulfate and nitrate concentration measurements. During the 2001 summer intensive field study chase studies on bus and truck emissions powered by Compressed Natural Gas (CNG), Diesel Filter Continuous Regenerating Technology (DF-CRT), and standard Heavy Duty diesel have been performed to characterize in-use emissions and the effectiveness of control technologies (CNG, DFCRT). The studies were performed in collaboration with NYC Transit Authority and NYS DEC and designed to specifically characterize the MTA bus fleet, which had introduced a fleet of CNG powered and DF-CRT retrofitted diesel buses. The study was to identify both the positive and potential negative impacts associated with the technologies. Results from these studies [Canagaratna, et al., 2004] showed significant reductions in in-use PM emissions from retrofitted DF-CRT buses, values quite consistent with those reported from chassis dynamometer studies. The studies also identified an NO_2 slip problem, where NO_2 used in the control process to react with carbon trapped on the oxidative catalytic surface is in excess under some load conditions. Although the net NO_x emission remains unchanged, the fraction NO_2/NO_x can be as high as 50% under some conditions. In addition, significant benefits in the use of low-sulfur fuel (required by the DF-CRT technology and used for the entire MTA fleet) have been observed as well as major PM emission reduction in CNG powered buses [Shorter et al, 2005; Herndon et al., 2005]. A major negative impact observed from monitoring in-use CNG powered buses was the emission of formaldehyde. The emissions of SO_2, H_2CO, and CH_4, measured using tunable infrared laser differential absorption spectroscopy, observed sulfur dioxide emissions from buses known to be burning ultra-low sulfur diesel (<30 ppm(m) sulfur) were 16 times lower than buses burning normal commercial diesel fuel, nominally less than 300 ppm(m) sulfur. Formaldehyde levels observed in sampled in-use CNG bus exhaust plumes were in the 100s of ppb range, levels typically a factor ~15 times greater than their diesel-powered counterparts. Such ambient hot spot concentrations are unacceptable and the recommendation has been made to introduce an oxidative catalysis on these exhaust systems, to reduce formaldehyde emissions from CNG powered vehicles. Finally, substantial work has been performed characterizing the chemical composition particulate matter at urban and regional sites in New York State [Drewnick et al., 2004a, b; Schwab et al., 2004a]. A combination of short term field intensive studies and routine measurements have provided detailed insights as to the chemical composition of PM as a function of size and the seasonal variation of PM2.5

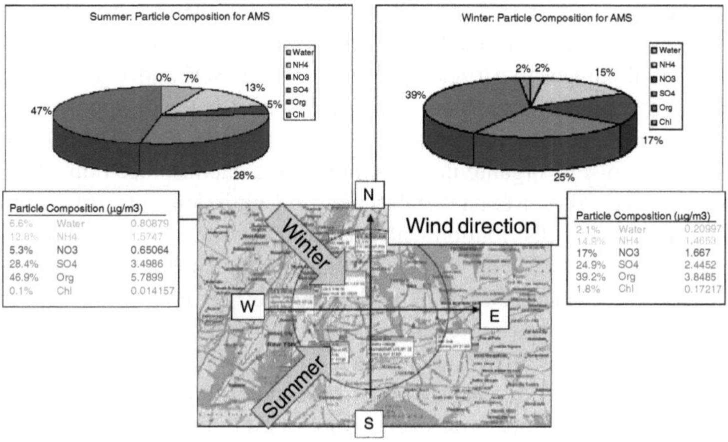

FIGURE 2. Summer 2001 vs. Winter 2004 Comparison of PM Composition in Queens New York.

composition across New York State. Figure 2 shows comparisons of mean PM composition from Aerosol Mass Spectrometer (AMS) measurements performed during PMTACS-NY 2001 Summer and 2004 Winter Intensive Field Studies. The results show somewhat higher PM sulfate and organic levels in the summer as compared to higher PM nitrate levels in the winter.

Comparisons of summer versus winter AMS PM size distributions measurements (Figure 3) show a distinct difference in the mean mode aerosol diameter,

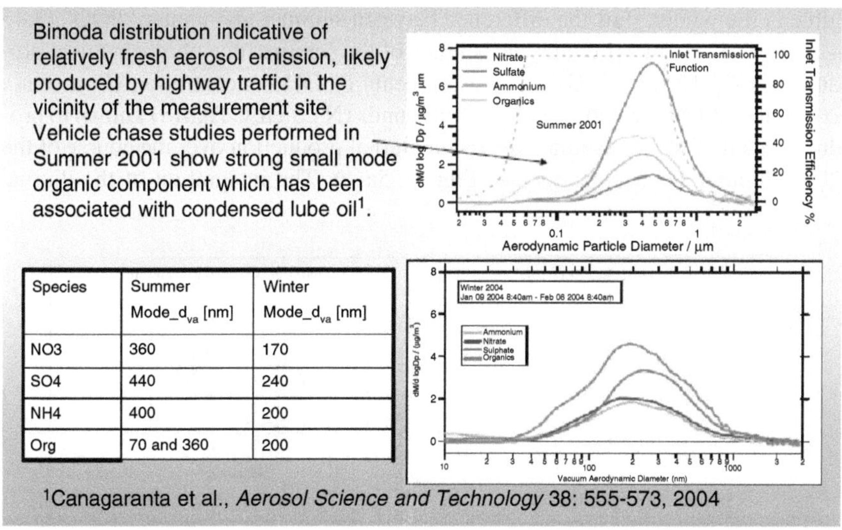

Species	Summer Mode_d_{va} [nm]	Winter Mode_d_{va} [nm]
NO3	360	170
SO4	440	240
NH4	400	200
Org	70 and 360	200

FIGURE 3. Summer 2001 versus Winter 2004 comparison of PM size distribution in Queens, New York.

with winter sizes smaller on average than summer sizes. We have evidence to suggest that this difference is likely due to secondary photochemical production processes involving gas to particle transformations of SO2 and VOCs which grow on existing aerosol surfaces. Also shown in Figure 3 is the disappearance of the bi-modal character of the PM organic in the wintertime data. The 70-nm mean mode diameter of PM organic in the summer-time data has been directly identified with primary diesel exhaust/lubricating oil emissions. A shifted in diesel exhaust emission size distribution with decreasing temperature has been observed by Abdul-Khalek et al., 1999, we believe that this and enhanced condensation on existing aerosol surfaces due to extreme cold weather conditions may explain these findings. We are currently studying this phenomenon in more detail as we reduce and analyze our aerosol sizing data.

Finally, we have investigated the role of secondary PM production from local summertime photochemical transformations of SO_2 to particulate sulfate and volatile organic compounds (VOCs) to secondary organic aerosol (SOA). We are particularly interested in an application of the AMS that looks extremely promising in providing estimates of SOA production and the role of VOC precursors in the attribution of PM organic mass in New York City. These results could have significant implications in the development of control strategies for both photochemical oxidants and PM. Using the monthly mean AMS data shown in Figure 4 we have estimate that 60% of the total summer organic (60% × 5.79 μg/m^3) = 3.47 μg/m^3 is attributable to "partially oxidized organics" ("POO"); while in the winter we estimate (35% × 3.85 μg/m^3 = 1.35 μg/m^3) is attributable to "POO" species.

Organics based on AMS mass fragments [(m43& m57) and m44, for Prim and POO respectively]. We can demonstrate that photochemical production is negligible in the winter, thus the difference between summer and winter "POO" (3.47 – 1.35 = 2.12 μg/m^3) provides an estimate of the summertime photochemical production of 2.12 μg/m^3. Using measurements of OH concentration (Ren et al., 2003) and selected volatile organic compounds (NYS DEC PAMS) known to produce organic PM, we estimate the mean diurnal production over the course of the 2001 summer field intensive (see Figures 5a–c). The integration of the diurnal

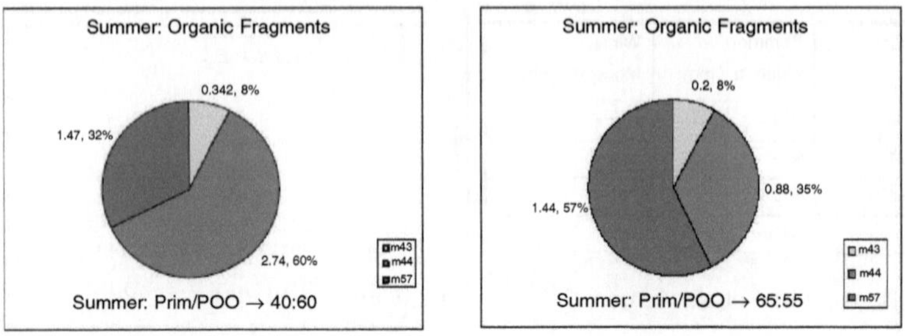

FIGURE 4. Summer 2001 versus Winter 2004 distribution of PM primary and partially oxidized.

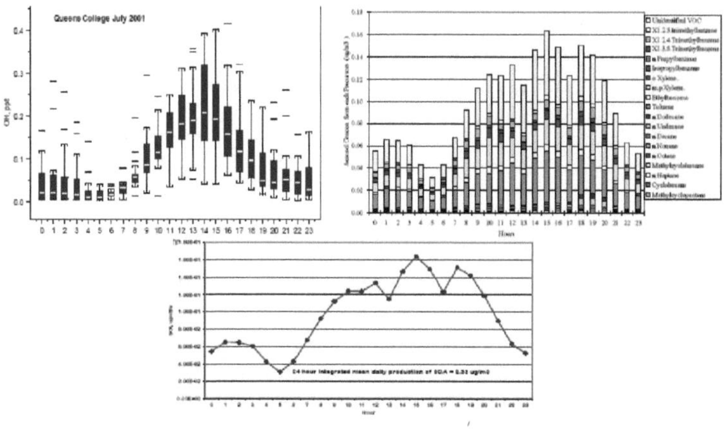

FIGURE 5A–C. Summer 2001 (a) mean diurnal OH measurements [based on Ren et al., 2003], (b) mean diurnal secondary PM organic production estimated from measurements of select VOCs and OH (c) mean diurnal integrated secondary PM organic production estimates in Queens, NY.

results gives a mean production of PM SOA for the campaign of 2.33μg/m^3, a production rate very consistent with that estimated from the AMS organics mass fragmentation analysis. Results from the U.S. EPA "Supersite" Programs have provided compelling evidence for the key PM compositional components contributing to nonattainment of the PM2.5 National Ambient Air Quality Standard (NAAQS). These results indicate for example in New York City that carbon- and sulfur-based PM contributed more than 70% of the total PM mass throughout the year, with PM nitrate of some significance only during the cold season. In preparation of state implementation plans for PM, most cities will have to assess control strategies that deal with precursor emissions (SO_2, VOC, and NO_x) and primary emissions of OC/EC to achieve the PM2.4 NAAQS.

Several pending national control programs slated for implementation in the next two years will likely be beneficial to the development of PM state implementation plans (SIPs). These included the new fuel sulfur rule and the 2007 heavy-duty diesel engine standards. But it is likely that additional controls will be needed. Preliminary results from Supersite field intensive studies in Queens, New York, suggest that an effective PM2.5 mitigation strategy would consider the implementation of a diesel filter trap catalytic converter retrofit control program for all heavy-duty diesel trucks operating in the major metropolitan areas have significant diesel truck traffic. Ultimately, whatever the strategy selected, it is critical to have measurements in place capable of tracking, through direct observation, changes in PM primary and precursor species concentrations in response to implemented control programs. This concept of *Accountability in Air Quality Management Systems* [Demerjian et al., 1995] is now accepted as a critical element in building credibility and public trust in our environmental regulations and their effectiveness in meeting air quality standards and anticipated improvements in environmental health. Continuation of "Supersite like" measurement programs is necessary to provide the essential monitoring data needed to track the progress of implemented air quality management strategies and addressing the reasons for their overall success or failure.

Reference

Abdul-Khalek, I., D. Kittelson, F. Bear, The influence of dilution conditions on diesel exhaust particle size distribution measurements, Society of Automotive Engineers, Paper No.1999-01-1142.

Albritton, D.L., D.S. Greenbaum, Atmospheric Observations: Helping Build the Scientific Basis for Decisions Related to Airborne Particulate Matter. Report of the PM Measurements Research Workshop, Chapel Hill, NC, 1998.

Canagaratna, M.J., J.T. Jayne, D. Ghertner, S. Herndon, Q. Shi, J.L. Jimenez, P.J. Silva, P. Williams, T. Lanni, F. Drewnick, K. L. Demerjian, C. Kolb, D. Worsnop, Chase studies of particulate emissions from in-use New York City vehicles, *Aerosol Science & Technology*, **38**, 555–573, 2004.

Demerjian, K.L., A review of National Monitoring Networks in North America, *Atmos. Environ*, **34**, 1861–1884, 2000.

Demerjian, K.L., P.M. Roth, C. Blanchard. A new approach for demonstrating attainment of the ambient ozone standard, EPA/600/R-96/134 , U.S. EPA ORD Research Triangle Park, NC 27711, October 1995.

Drewnick, F., J.J. Schwab, O. Hogrefe, S. Peters, L. Husain, D. Diamond, R. Weber, K.L. Demerjian, Intercomparison and evaluation of four semi-continuous PM2.5 sulfate instruments, *Atmos Environ.* **37**, 3335–3350, 2003.

Drewnick, F., J.J. Schwab, J.T. Jayne, M. Canagaratna, D.R. Worsnop, K.L. Demerjian, Measurement of ambient aerosol composition during the PMTACS-NY 2001 campaign using an aerosol mass spectrometer. Part I: Mass concentrations, *Aerosol Science and Technology*, **38**(SI), 92–103, 2004a.

Drewnick, F., J.T. Jayne, M. Canagaratna, D.R. Worsnop, K.L. Demerjian, Measurement of ambient aerosol composition during the PMTACS-NY 2001 campaign using an aerosol mass spectrometer. Part II: Chemically speciated mass distribution, *Aerosol Science and Technology*, **38**(SI):104–117, 2004b.

Dutkiewicz, V.A., S. Qureshi, A.R. Khan, V. Ferrara, J. Schwab, K. Demerjian, L. Husain, Sources of fine particulate sulfate in New York. *Atmos. Environ.*, **38**, 3179–3189, 2004.

Hering, S., P.M. Fine, C. Sioutas, P.A. Jacques, J.L. Ambs, O. Hogrefe, K.L. Demerjian, Field assessment of the dynamics of the particulate nitrate vaporization using differential TEOM andautomated nitrate monitors, *Atmos. Environ.*, **38**, 5183–5192, 2004.

Herndon, S.C., J.H. Shorter, M.S. Zahniser, J. Wormhoudt, D.D. Nelson, K.L. Demerjian, C.E. Kolb, Real-time measurements of SO_2, H_2CO and CH_4 emissions from in-use curbside passenger buses in New York City using a chase vehicle" (in review ES&T).

Hogrefe, O., F. Drewnick, G.G. Lala, J. J. Schwab, K.L. Demerjian, Development, operation and applications of an aerosol generation, calibration and research facility, new instruments and data inversion methods, *Aerosol Science and Technology*, **38**(SI), 196–214, 2004.

Hogrefe, O., J. Schwab, F. Drewnick, K. Rhoads, G.G. Lala, H.D. Felton, O.V. Rattigan, L. Husain, V.A. Dutkiewicz, S. Peters, K.L. Demerjian, Semi-continuous PM2.5 sulfate and nitrate measurements at an urban and a rural location in New York: PMTACS-NY Summer 2001 and 2002 campaigns., *J. Air & Waste Manage. Assoc.*, **54**, 1040–1060, 2004.

Jayne, J.T. et al., Development of an Aerosol Mass Spectrometer for Size and Composition Analysis of Submicron Particles, *Aerosol Sci. Technol.*, in press, 1999.

Li, Z., P.K. Hopke, L. Husain, S. Qureshi, V.A. Dutkiewicz, J.J. Schwab, F. Drewnick, K.L. Demerjian, Sources of fine particle composition in New York City, *Atmospheric Environment*, **38**, 6521–6529, 2004.

Li, Y.Q., K.L. Demerjian, M.S. Zahniser, D.D. Nelson, J.B. McManus, S.C. Herndon, Measurements of formaldehyde, nitrogen dioxide, and sulfur dioxide at Whiteface Mountain using a dual tunable diode laser system, *J. Geophys. Res.* **109**, D16S08, doi:10.1029/2003JD004091, 2004. NARSTO (2003). Particulate Matter Science for Policy Makers -A NARSTOAssessment, February 2003. http://www.cgenv.com/Narsto/.

Ren, X., H. Harder, M. Martinez, R.L. Lesher, A. Oliger, T. Shirley, J. Adams, J. B. Simpas, W.H. Brune, J, HOx concentrations and OH reactivity observations in New York City during PMTACS-NY 2001, *Atmospheric Environment*, **37**, 3627–3637, 2003.

Ren, X., H. Harder, M. Martinez, R.L. Lesher, A. Oliger, J. B. Simpas, W.H. Brune, J. J. Schwab, K. L. Demerjian, Y. He, X. Zhou, H. Gao, OH and HO_2 chemistry in the urban atmosphere of New York City, *Atmospheric Environment*, **37**(26), 3627–3637, 2003.

K.L. Demerjian, Y. He, X. Zhou, H. Gao, OH and HO_2 chemistry in the urban atmosphere of New York City, *Atmospheric Environment*, **37**, 3639–3651, 2003.

Schwab, J.J., J. Spicer, H.D. Felton, J.A. Ambs, K.L. Demerjian, Long-term comparison of TEOM, SES TEOM and FRM measurements at rural and urban New York sites, in Symposium on Air Quality Measurements and Technology – *2002I*, VIP-115-CD, *ISBN 0-923204-50-4*, Air and Waste Management Association, Pittsburgh, PA, 2003.

Schwab, J.J., H.D. Felton, K.L. Demerjian, Aerosol chemical composition in New York state from integrated filter samples: Urban/rural and seasonal contrasts. *J. Geophys. Res.*, **109**, D16S05, doi:10.1029/2003JD004078, 2004a.

Schwab, J.J., J. Spicer, K.L. Demerjian, J.L. Ambs, H.D. Felton, Long-term field characterization of TEOM and modified TEOM samplers in urban and rural New York State locations, *J. Air & Waste Manage.*, **54**, 1264–1280, 2004b.

Schwab, J.J., O. Hogrefe, K.L. Demerjian, J.L. Ambs, Laboratory characterization of modified TEOM samplers, *J. Air & Waste Manage.*, **54**, 1254–1263, 2004c.

Schwab, J.J., Y.-Q. Li, K.L. Demerjian, Semi-continuous formaldehyde measurements with a diffusion scrubber/liquid fluorescence analyzer. In Symposium on air quality measurement methods and technology—2004 [CD-ROM], Air and Waste Management Association, Pittsburgh, PA, 2004d.

Shorter, J.H., S.C. Herndon, M. S. Zahniser, D. D. Nelson, J. Wormhoudt, K. L. Demerjian, C.E. Kolb, Real-time measurements of nitrogen oxide emissions from in-use New York City transit buses using a chase vehicle (in review ES&T). U.S. EPA 2004, The National Ambient Air Monitoring Strategy, Office of Air Quality Planning and Standards, Research Triangle Park, NC, April 2004.

Yu, F., T. Lanni, B. Frank, Measurements of ion concentration in gasoline and diesel engine exhaust, *Atmospheric Environment*, **38**, 10, 1417–1423, 2003.

Zhou, X., H. Gao, Y. He, G. Hung, S.B. Bertman, K. Civerolo, J. Schwab, Nitric acid photolysis on surfaces in low-NO_x environments: Significant atmospheric implications. *Geophys. Res. Lett.*, **30**, 2217, doi:10.1029/2003GL108620, 2003.

Part III
Ground-Based Networks

Chapter 10

Ozone from Soundings: A Vital Element of Regional and Global Measurement Strategies

ANNE M. THOMPSON

Abstract. Satellite instruments measuring ozone, predictive models of atmospheric trace gases, and models employing assimilation require well-resolved, accurate, and precise ozone profiles in the stratosphere and troposphere. These are provided by the ozonesonde instrument, typically an electrochemical concentration cell device flown on a balloon with a standard radiosonde. The design elements of a successful network of ozonesondes in the tropics (SHADOZ = Southern Hemisphere Additional Ozonesondes; Thompson et al., 2003a,b) are described. Sondes are a vital component of a sensor web as described by Schoeberl and Talabac [this volume] because profile variability can optimize retrieval accuracy as air mass types change. More generally, combined with assimilation, profiles from sondes and aircraft, as well as ground-based measurements are essential for accurate global measurement of tropospheric ozone.

Introduction

Tropospheric ozone is of great interest locally as a pollutant. Ambient ozone concentrations greater than about 100 ppbv (parts-per-billion by volume) are deemed unhealthy and in many places emissions of the "precursor" gases (hydrocarbons, carbon monoxide, nitrogen oxides; for reactions, refer to Brune [this volume] leading to ozone formation are regulated. Ozone is measured continuously at the surface in hundreds of cities around the world. Globally, our concerns about ozone are its role in transboundary pollution, from city to city, region to region, and across continents and oceans. In addition, potential increases in the free tropospheric concentration of ozone are of interest because tropospheric ozone is a greenhouse gas. For global measurement, satellite instrumentation, aircraft, and sondes are employed. For example, GOME (acronyms given at end of text) and SCIAMACHY directly, and TOMS indirectly, are able to track ozone pollution with some resolution in the free troposphere. Aura, launched in July 2004, and presently in the testing phase, will expand tropospheric ozone measurements. Present satellite technology, however, does not resolve the vertical ozone distribution required for following trends [WMO, 1998] and for evaluating and constraining models. Thus, vertical profiling of ozone from aircraft [Cammas, this volume] and from balloon-borne ozonesondes, as presented in this chapter, are required.

The roles of aircraft and sonde profiles in a sensor-web approach [Schoeberl and Talabac, this volume] to ozone measurement are complementary. Aircraft can operate regularly, taking frequent profiles with automated instrumentation over many locations, but within an altitude range limited to about 12 km [Marenco et al., 1998; Thouret et al., 1998]. Sondes are more labor-intensive, relatively expensive, and therefore are usually launched at weekly-monthly frequency [WMO, 1998] at a given site. However, in contrast to aircraft sampling, both tropospheric and stratospheric ozone (to 5–10 hPa) are sampled at 10–15 m resolution as the sonde ascends. Below, the rationale for ozone measurement by sondes and characteristics of ozone profiles are summarized. Then the design of a paradigmatic ozonesonde network for specific scientific objectives is described. Finally, the incorporation of ozonesondes in a sensor web that makes use of new satellite instruments and models is described.

Ozone Measurement Requirements and Profiles

There are three major pollutants presently measured by satellite—ozone, carbon monoxide, and aerosols—all with lifetimes of days-months (Figure 2 of Brune, this volume, where PM 2.5 denotes aerosols). From aircraft, ground-based (e.g., lidar) and sounding observations, it is well-known that thin layers of the three pollutants can be very stable, making profiles highly variable. An example from an ozone sounding is shown in Figure 1 and the sonde instrument is shown in Figure 2. A set of tropical profiles from Ascension Island in the south Atlantic (8S, 14.4W), where ozone is characterized by low surface values (< 40 ppbv, purple trace at right) is illustrated. Aloft is pollution (layers with > 100 ppbv), and a distinct tropopause, signified by a rapid ozone increase and temperature inversion (red) at ~ 100 mb. Air parcel back-trajectories from pollution layers in the sounding show origins over Africa where biomass fires are active in September. See trajectories under "Ascension" and the corresponding year and date at: http://croc.gsfc.nasa.gov/shadoz

The column-integrated total ozone amount for the Ascension sounding, with extrapolation based on the SBUV satellite profile climatology above 7 mb (59 DU, Dobson Units; one DU = 2.69×10^{16} molec/cm[2]), is 287 DU.

Design of a Tropical Ozonesonde Network and Key Findings

The principles of an observing system incorporating ozonesondes are illustrated by a pair of networks we have designed. The discussion here is based on the SHADOZ (Southern Hemisphere Additional Ozonesondes) network we have organized in the southern tropics and subtropics. Since its inception in 1998, over

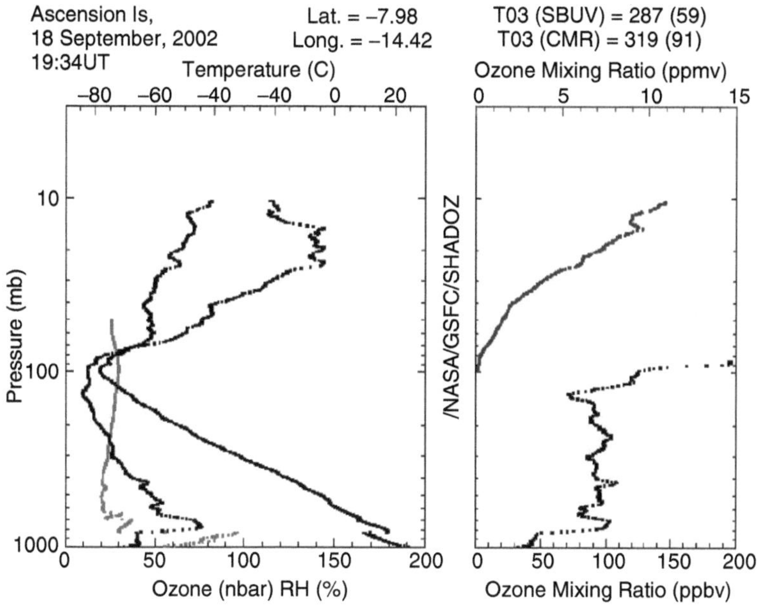

FIGURE 1. Ozone (partial pressure) and radio sounding (temperature and relative humidity change with pressure) data from launch at Ascension Island (18 October 2002) on left. On the right side, mixing ratio of ozone is displayed in ppbv (parts-per-billion by volume) to 100 hPa and in ppmv (parts-per-million by volume) in the stratosphere. From SHADOZ (Southern Hemisphere Additional Ozonesondes) Web site: <http://croc.gsfc.nasa.gov/shadoz>.

2300 ozone (and pressure-temperature-humidity, PTU) profiles have been archived at the SHADOZ Web site: http://croc.gsfc.nasa.gov/shadoz. The requirements and responsive design features of the network are summarized along with illustrations of scientific findings. In addition, we have found gaps in the present network and so we mention needs that new stations (and complementary approaches, e.g., aircraft profiling) can fulfill.

The questions that SHADOZ addresses are as follows:

- Are tropospheric ozone satellite columns derived from satellite instruments (with TOMS, GOME, SCIAMACHY, and other instruments) accurate? *Profile measurements are needed at sub-km resolution throughout troposphere and stratosphere. Only soundings meet this need with consistent sampling quality.*
- What causes the zonal wave-one pattern observed in total equatorial ozone [Fishman and Larsen, 1987; Shiotani, 1992] and described further by Hudson and Thompson [1998] *The geographic distribution of sounding sites must span the full tropical longitude range.*
- What is the regional and temporal variability (over days-to-season scale) of tropical ozone? Are meteorological or photochemical factors primarily responsible

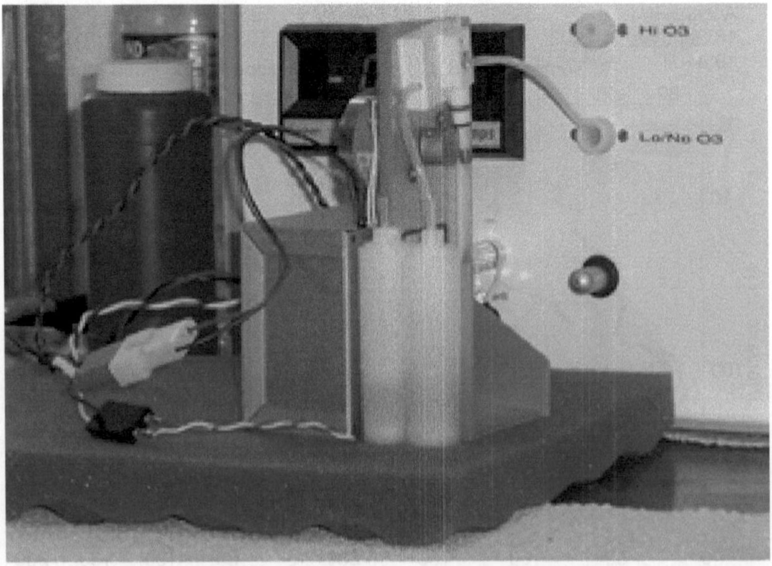

FIGURE 2. Closeup of ozonesonde of electrochemical concentration cell (ECC) type.

for observed ozone variations? *Weekly sampling is nominal; a few stations launch twice per month.* Note that practical aspects of the ozonesonde measurement were factors determining SHADOZ site selection and sampling frequency. First, operational stations were chosen. Second, requirements of personnel and cost for ozonesonde operation (500–700 USD for instruments, balloon inflation gas and other supplies at each launch) and chemical preparation, launch and processing time (two days each) imposed practical limits.

The SHADOZ network (as of 2004) is illustrated in Figure 3. Irene (late 1998), Malindi and Paramaribo (1999), Kuala Lumpur (data available in archive from 1998) joined after the 1998 initiation. Details of site location, operations and instrumentation appear in Thompson et al. [2003a; 2004].

Findings from SHADOZ, in response to the questions above, are as follows.

- Comparisons of total and tropospheric ozone from SHADOZ with TOMS satellite measurements appear in Thompson et al. [2003a,b] for 1998-2000. In general, TOMS total ozone is greater than the sonde value (example in Figure 4). In some cases, the sonde probably measures too low in the stratosphere [Smit and Sträter, 2004]. In other cases (as for Samoa, in Figure 4), the TOMS algorithm is known to overestimate tropospheric ozone (Figure 8 in Thompson et al., 2003b). A number of tropospheric ozone product comparisons have evolved for which SHADOZ comparisons are being made [Ziemke et al., 2002; Kim et al., 2004].
- The wave-one feature that guided SHADOZ design, as it appears schematically in Figure 5, refers to a 15–20 DU higher total ozone column amount

FIGURE 3. Map of SHADOZ stations current to 2004. See Thompson et al. [2003a, 2004] for latitude and longitude of each site. Operations at Tahiti ended after 1999. These stations began operating after the January 1998 start of SHADOZ: Irene (South Africa); Malindi (Kenya); Paramaribo (Surinam).

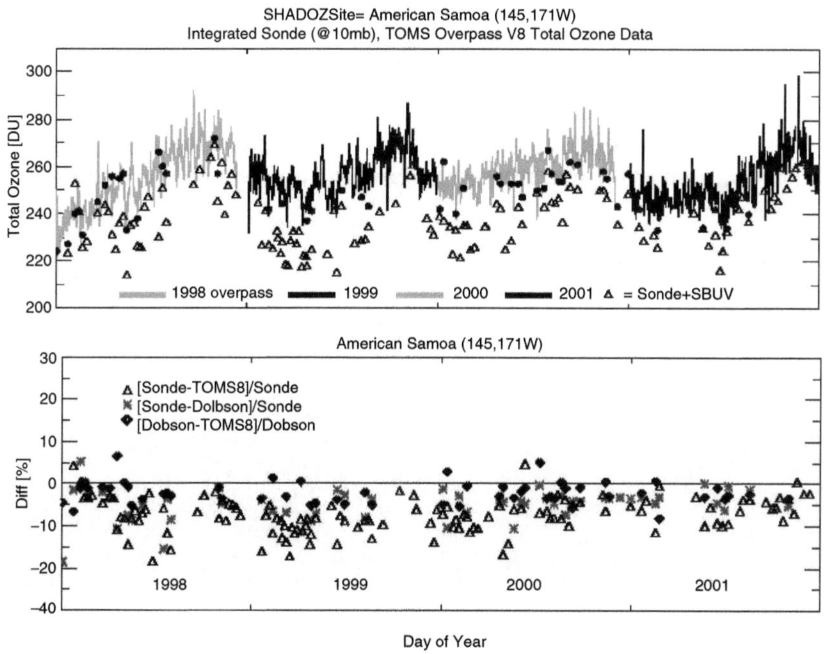

FIGURE 4. Comparison of daily TOMS (version 8) ozone data from the satellite measuring as it passed over Samoa (14S, 171W), along with the Dobson total ozone measured from the surface at Samoa and total ozone integrated from the sondes launched in 1998–2001. Disagreement between the sonde and TOMS and the Dobson and TOMS (the satellite is typically higher in total ozone) is due to two factors: the satellite algorithm is artificially high in the troposphere; the particular sonde variety used in the measurement is apparently biased low in parts of the stratosphere [Thompson et al., 2003a; Smit and Sträter, 2004].

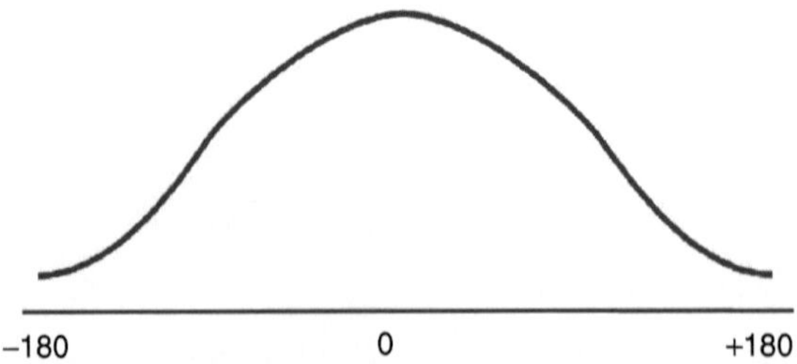

FIGURE 5. Schematic representation of the wave-one feature in total ozone that is observed in the TOMS total ozone measurement and in total ozone integrated from SHADOZ sondes. Longitude is the horizontal axis. Exact location of maximum and minimum vary, as does the magnitude of highest ozone (5–20 Dobson Units).

over the Atlantic-Africa-western Indian Ocean region than over the mid-Pacific. Analysis of the cross-sectional structure of ozone reveals that the integrated stratospheric column amount is invariant with longitude, to within the accuracy of the sonde technique. The tropospheric ozone column varies, however, with the same magnitude as found by the satellite measurement. Figure 6 (adapted from Thompson et al., 2003b, 2004), based on all the southern hemisphere SHADOZ data except for Irene, shows that more ozone tends

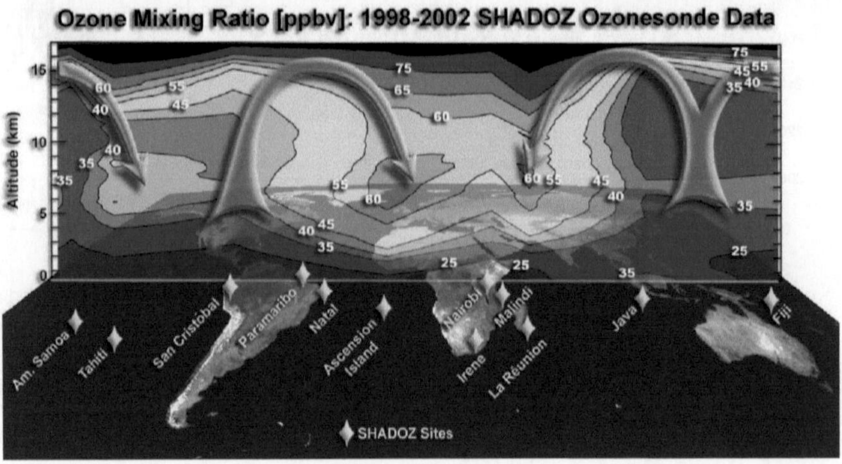

FIGURE 6. Cross-section of ozone mixing ratio based on 1998–2002 data averaged at 0.25 km intervals from the stations illustrated (except Irene and Paramaribo). More ozone is observed over the region spanned by eastern South America-Atlantic-Africa-western Indian Ocean than over the Pacific and eastern Indian Ocean. When integrated tropospheric ozone is compared between Atlantic and Pacific, the difference is equal to the typical "wave-one" magnitude. Thus, it is seen that the distribution of tropospheric ozone is responsible for the wave-one pattern. Stratospheric ozone, in contrast, is uniform over the same region (within experimental error; see Thompson et al., 2003a; Figures 12–14).

to be concentrated in the free troposphere in the Atlantic-Africa-western Indian Ocean region. The causes of higher ozone in the region of the maximum include general circulation (subsidence indicated by downward arrows), a lower tropopause, and a greater concentration of ozone chemical precursors (from lightning and combustion) in the 40W to 60E region, compared to the Pacific [Oltmans et al., 2001].

- Analysis of temporal variability at a single SHADOZ station (e.g., Figure 9 in Thompson et al., 2003b) shows signals with a 3–7-day period. For example, this type of periodicity has been deduced from Nairobi record (Figure 7). Unexpected regional variability emerged in several studies. For example, in early 1999, during the Aerosols99 trans-Atlantic ship crossing [Thompson et al., 2000] and the time of the INDOEX experiment [Chatfield et al., 2004], higher ozone in the troposphere appeared over the southern tropical Atlantic than at the same latitudes north of the ITCZ where biomass burning was active. This phenomenon is referred to as the "tropical Atlantic paradox" (depicted in ozone latitudinal mixing ratio cross-section in Figure 8). When ozone seasonality is compared at Paramaribo (Figure 8) to that for a southern hemisphere SHADOZ site (see Figure 8 in Thompson, 2004), there is a contrast, primarily in the January-to-March period. This is due to different "seasonality" at Surinam in terms of ITCZ position and precipitation and winds compared to the southern tropical stations [Peters et al., 2004].

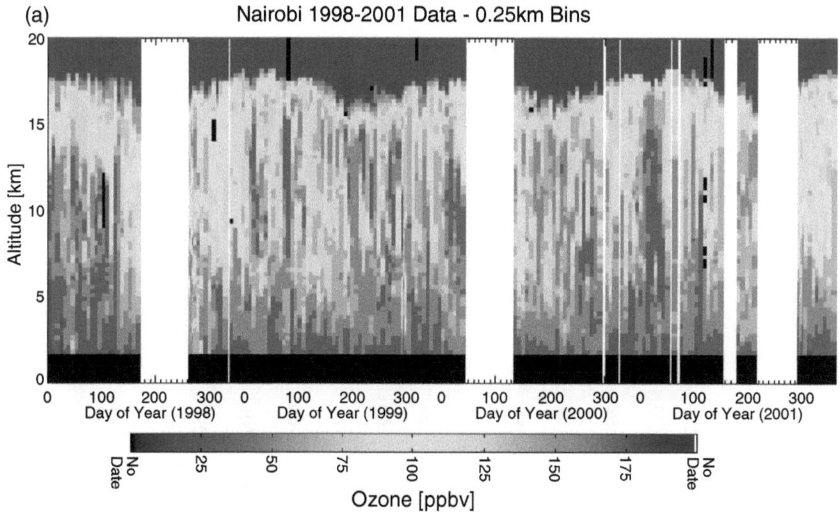

FIGURE 7. Ozone mixing ratios, 0.25-km means, from weekly soundings over Nairobi from 1998–2001 (adapted from Thompson et al., 2004). Time-series analysis shows cycles on an annual basis, semi-annual, 30–60 days (the frequency of the Madden–Julian Oscillation) and a signal less than 10 days. Advective and convective processes, introducing cleaner or more polluted air masses to the Nairobi area, are responsible for much of the short-term variability [Chatfield et al., 2004].

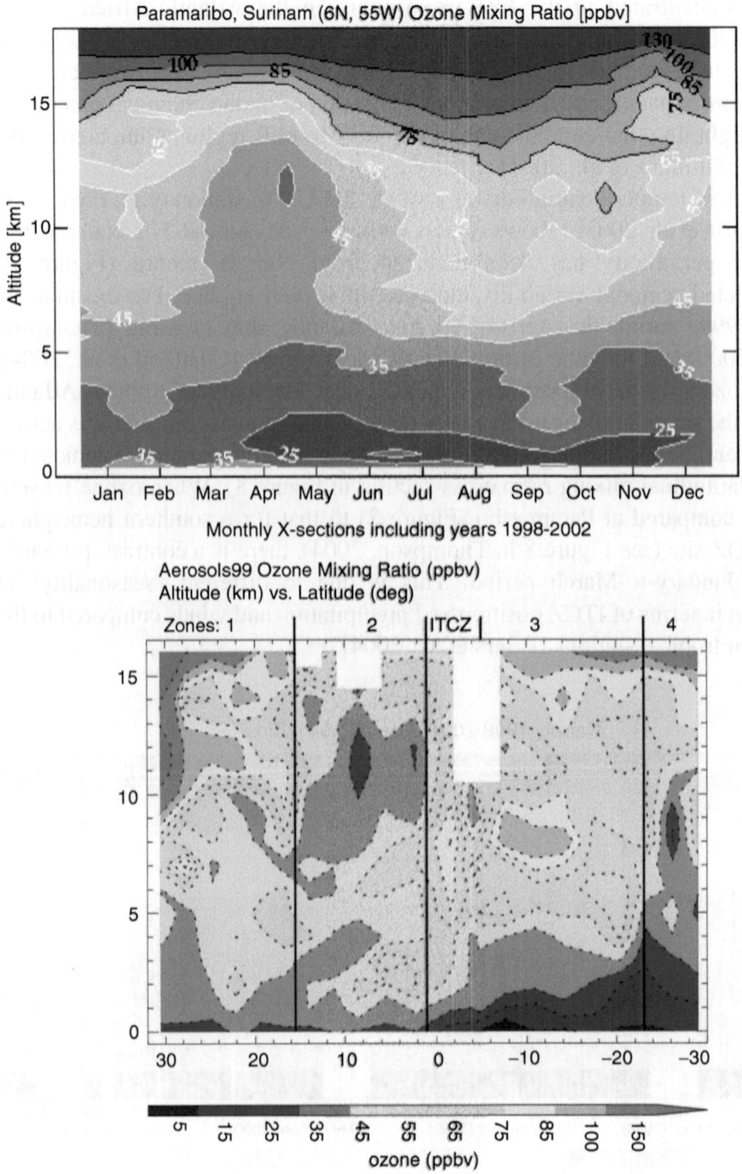

FIGURE 8. (A) Latitudinal ozone cross-section below 17 km measured by soundings during the Aerosols99 trans-Atlantic cruise (figure adapted from [Thompson et al., 2000]). The higher ozone in the lower and middle troposphere over the north tropical Atlantic (designated "2") in the figure is due to pollution from biomass fires over north Africa. However, in zone "3", south of the equator where the wet season is prevalent, integrated tropospheric ozone is higher, due to the concentrated higher-ozone amounts in the middle-upper troposphere. (B) Seasonal variation in tropospheric ozone over Paramaribo, based on soundings averaged over each month during 1999–2003.

- Another intriguing variability in ozone regional distributions has been described by Sauvage et al. [2004]. The ozone distribution based on the MOZAIC [Marenco et al., 1998; Thouret et al., 1998] profiles from aircraft landings and takeoffs over Africa during December-January-February shows a latitudinal distribution without the "paradox" feature. As one would expect from proximity to biomass fires, there is higher ozone in the lower troposphere north of the equator (integrating to greater total tropospheric ozone) than there is in the southern hemisphere. Relatively enhanced upper tropospheric ozone (similar to that shown in Figure 8) is observed in the aircraft profiles over the African continent, similar to the Atlantic pattern, and consistent with the large-scale feature of subsidence depicted in Figure 6.

These discoveries of contrast between northern and southern hemispheric ozone and the apparent lack of a "paradox" over the African continent are leading to a revised design for SHADOZ as we move to the Aura era [Bhartia, this volume]. Potential new regions for SHADOZ sites are in west Africa, southern Asia, the middle East, central America. In addition, interferences of aerosols with ozone retrieval cause us to consider more pollution as a criterion for new network locations.

The Ozonesonde as an Essential Element of Integrated Atmospheric Chemical Observations

Figure 9 shows schematically that ozonesondes can integrate observations and assimilation model representations of ozone distribution in the atmosphere. Satellites are not yet able to resolve ozone concentrations in the vertical to the

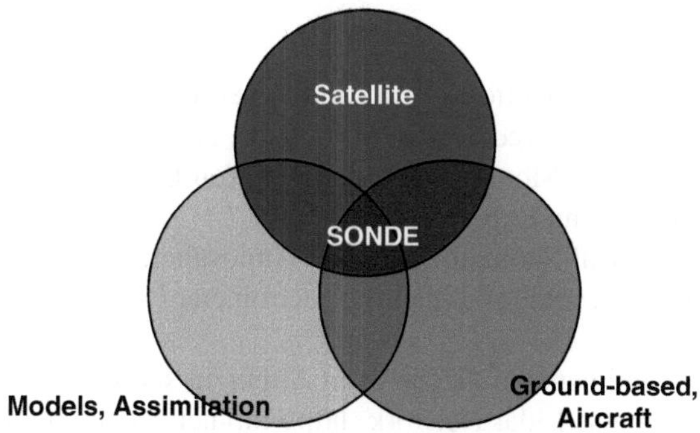

FIGURE 9. Schematic of sondes as an integrating element of information among models, aircraft profiles, satellite column measurements for the derivation of tropospheric ozone. Neither models nor satellite are able to consistently derive the location of the tropopause and aircraft profiling encompasses this boundary only when it is below about 200 hPa. Sondes provide locate the tropopause region with the resolution required for guiding satellite algorithms and corresponding assimilation.

extent required for tracking pollution transport. Nor do they locate the tropopause reliably enough for separation of stratospheric and tropospheric ozone [Thompson et al., 2001]. Sondes can do this consistently but their geographical and temporal coverage is limited. Thus, multiple observing elements need to be present in order to obtain regional and global three-dimensional ozone distributions.

There are several ways that sondes can be used to optimize ozone retrieval from satellite. One is that buv instruments (e.g., TOMS, OMI) start with first-guess profiles in which stratospheric ozone is varied according to total ozone amount and latitude. Ozonesonde climatologies like SHADOZ have shown that latitude alone (and season) are inadequate for specifying a first-guess profile (resulting in the disagreement shown in Figure 4). However, sonde climatologies can be prepared according to total ozone [Hudson et al., 2003] with statistically derived corresponding stratospheric and tropospheric ozone profiles. Models have evolved to give quite accurate tropopause forecasts. This tropopause information can be combined with total ozone and first-guess tropospheric and stratospheric profiles for a given region and tropopause location to derive improved ozone column and profile amounts. Another approach that would make use of sondes is to combine them with multiple satellite information and cloud and aerosol data. The latter can originate from aerosol and cloud sensors aboard Aqua and correlative ground-based instrumentation. Possibilities include MODIS, refined with observations from AERONET and MPL-Net, and aerosol models.

Acknowledgements The author thanks J. C. Witte (SSAI at NASA Goddard Space Flight Center) for graphical analysis and J. B. Stone (Penn State University) for assistance with the manuscript. The author is grateful to W. H. Brune (Penn State University) and P. Di Carlo (University of L'Aquila) for the opportunity to participate in ISSAOS, September 2004.

Acronyms

AERONET = Aerosol Robotic Network (http://aeronet.gsfc.nasa.gov)
AIRS = Atmospheric Infrared Sounder (on Aqua, 2002-)
GOME = Global Ozone Monitoring Experiment (on ERS-1, 1995-)
INDOEX = Indian Ocean Experiment (1999)
ISSAOS = International Summer School on Atmospheric and Oceanic Science
MODIS = Moderate-resolution Imaging Spectrometer (on Terra, 1999-; on Aqua, 2002)
MOZAIC = Measurements of Ozone from Airbus In-service Aircraft (1994-)
MPL-Net = Micropulse Lidar Network (http://mplnet.gsfc.nasa.gov)
OMI = Ozone Monitoring Instrument (on Aura, 2004-)
SCIAMACHY = Scanning Imaging Absorption Spectro-Meter for Atmospheric Charto-graphY
SHADOZ = Southern Hemisphere Additional Ozonesondes
TES = Tropospheric Emission Spectrometer (on Aura, 2004-)
TOMS = Total Ozone Mapping Spectrometer (on Earth-Probe, 1996-)

References

Brune, W.H., this volume, 2006.

Cammas, J-P., this volume, 2006.

Chatfield, R.B., H. Guan, A.M. Thompson, J.C. Witte, Convective lofting links Indian Ocean air pollution to paradoxical south Atlantic ozone maxima, *Geophys. Res. Lett.*, **31**, L06103, doi: 10.129/2003GL018866, 2004.

Fishman, J., J.C. Larsen, Distribution of total ozone and stratospheric ozone in the tropics -Implications for the distribution of tropospheric ozone, *J. Geophys. Res.*, **92**, 6627–6634, 1987.

Hudson, R.D., A.D. Frolov, M.F. Andrade, M.B. Follette, The total ozone field separated into meteorological regimes. Part I: Defining the regimes, J. Geophys. Res., **60**(14), 1669–1677, 2003.

Hudson, R.D., A.M. Thompson, Tropical Tropospheric Ozone (TTO) maps from TOMS by a modified residual method, *J. Geophys. Res.*, **103**, 22,129–22,145, 1998.

Kim, J.H., S. Na, M. Newchurch, K.J. Ha, Comparison of scan-angle method and convective cloud differential method in retrieving tropospheric ozone from TOMS, *Environmental Monitoring and Assessment*, **92**, 25–33, 2004.

IGOS Integrated Global Observing Strategy and the IGOS partnership (report) (http://ioc.unesco.org/igospartners/).

Marenco, A., V. Thouret, P. Nedelec, H. Smit, M. Helten, D. Kley, F. Karcher, P. Simon, K. Law, J. Pyle, G. Poschmann, R. Von Wrede, C. Hume, T. Cook, Measurements of ozone and water vapor by Airbus in-service aircraft: The MOZAIC airborne program, An overview, *J. Geophys. Res*, **103**(D19), 25631–25642, 10.1029/98JD00977, 1998.

Oltmans, S.J. et al., Ozone in the Pacific tropical troposphere from ozonesonde observations, *J. Geophys. Res.*, **106**, 32503–32526, 2001.

Peters, W., M.C. Krol, J.P.F. Fortuin, H.M. Kelder, C.R. Becker, A.M. Thompson, J. Lelieveld, P. J. Crutzen, Tropospheric ozone over a tropical Atlantic station in the northern hemisphere: Paramaribo, Surinam (6N, 55W), *Tellus B,* **56**, 21–34, 2004.

Randriambelo, T., J-L. Baray, S. Baldy, A.M. Thompson, S.J. Oltmans, P. Keckhut, Investigation of the short-term variability of tropical tropospheric ozone, *Annales Geophysiques,* **21**, 2095–2106, 2003.

Sauvage, B., V. Thouret, J-P. Cammas, F. Gueusi, G. Athier and P. Nédélec, Tropospheric ozone over Equatorial Africa: Regional aspects from the MOZAIC data, *Atmos. Chem. Phys. Disc.*, SRef-ID: 1680-7375/acpd/2004-4-3285, 2004.

Schoeberl, M.R. and Talabec, this volume, 2006.

Shiotani, M. Annual, quasi-biennial and El Nino-Southern Oscillation (ENSO) time-scale variations in Equatorial total ozone, *J. Geophys. Res.*, **97**, 7625–7634, 1992.

Smit, H.G.J., W. Sträter, JOSIE-1998: Performance of ECC Ozone Sondes of SPC-6A and ENSCI-A Type, in *WMO Global Atmospheric Watch Report Series, No. 157 (Technical Document No. 1218), World Meteorological Organization,* Geneva, 2004a.

Smit, H.G.J., W. Sträter, JOSIE-2000: The 2000 WMO international intercomparison of operating procedures for ECC-sondes at the environmental simulation facility at Jülich, in *WMO Global Atmospheric Watch Report Series (Technical Document)*, WMO Global Atmosphere Watch report series, No. 157 (Technical Document No. 1218), World Meteorological Organization, Geneva, 2004b.

Thompson, A.M., et al. A tropical Atlantic ozone paradox: Shipboard and satellite views of a tropospheric ozone maximum and wave-one in January-February 1999, *Geophys. Res., Lett.*, **27**, 3317–3320, 2000.

Thompson, A.M., R.D. Hudson, A.D. Frolov, J.C. Witte, T.L. Kucsera, Tropospheric ozone from space: Tracking pollution with the TOMS (Total Ozone Mapping Spectrometer) instrument, IGARSS Proceedings, Sydney, Australia, Meeting, July 2001, IEEE Publ., Piscataway, NJ, 2001.

Thompson, A.M., et al., Southern Hemisphere Additional Ozonesondes (SHADOZ) 1998–2000 tropical ozone climatology: 1. Comparison with TOMS and ground-based measurements, *J. Geophys. Res.*, **108**, 8238, doi: 10.129/2001JD000967, 2003a.

Thompson, A.M., et al., Southern Hemisphere Additional Ozonesondes (SHADOZ) 1998-2000 tropical ozone climatology. 2. Tropospheric variability and the zonal wave-one, *J. Geophys. Res.*, **108**, 8241, doi: 10.129/2002JD002241, 2003b.

Thompson, A.M., J.C. Witte, S.J. Oltmans, F.J. Schmidlin, SHADOZ (Southern Hemisphere Additional Ozonesondes): A tropical ozonesonde-radiosonde network for the atmospheric community, *Bull. Am. Meteorol. Soc.,* **85**, 1549–1564, 2004.

Thouret, V., A. Marenco, J.A. Logan, P. Nedelec, C. Grouhel, Comparisons of ozone measurements from the MOZAIC airborne program and the ozone sounding network at eight locations, *J Geophys. Res.*, **103**(D19), 25685–25720, 10.1029/98JD02243, 1998.

WMO (World Meteorological Organization), SPARC/IOC/GAW Assessment of Trends in the Vertical Distribution of Ozone, ed. By N. Harris, R. Hudson and C. Phillips, SPARC Report No. 1, WMO Global Ozone Research and Monitoring Project, Report No. 43., 1998.

Ziemke, J.R., S. Chandra, P.K. Bhartia, "Cloud Slicing": A new technique to derive upper tropospheric ozone from satellite measurements, *J. Geophys. Res.*, **106** (D9), 9853–9868, 10.1029/2000JD900768, 2001.

Chapter 11

Lidar Networks

VINCENZO RIZI

Introduction

This lecture covers some aspects of the lidar networks as systems for the observation of the atmosphere. After a short discussion of the lidar technique and of the typical lidar products, a few words are devoted to a partial list of the recent *lidar network* experiments. Finally, a discussion of the scientific problems and studies that can be addressed within a coordinated lidar network is presented (the peculiar example is a Raman-based lidar network, EARLINET, supported in 2000-2003 period by European Community).

The LIDAR (LIght Detection And Ranging) technique is quite diffuse among atmospheric studies. Figure 1 shows an elementary representation of this remote probing and sampling (with light as information carrier) of the atmosphere.

The laser emitted photons, while traveling in the atmosphere, are scattered, by different processes, of different efficiencies, involving the laser light features and the different components of the atmosphere.

The backscattered photons are collected by a telescope and detected by a configuration of photo-detectors as photomultipliers, etc.. Collecting the backscattered photons (not necessary at the same wavelength of the emitted ones) as a function of their flight time, it is possible to discriminate the photons returning from different ranges/altitudes. The number of collected photons depends of the atmospheric optical transmission (along the upward and downward travel), which origininates from the scattering and absorption processes, and of the local backscattering process.

$$L^{\lambda}(s) \propto L_o^{\lambda_o} \cdot T_{up}^{\lambda_o}(s) \cdot \left[\text{backscattering}(s)\right] \cdot T_{down}^{\lambda}(s)\,\frac{d\Omega}{4\pi} \qquad (1)$$

$L^{\lambda}(s)$ is the number of collected photons from the range s; $L_o^{\lambda_o}$ is the number of photons emitted by the laser; [backscattering(s)] includes all the effective scattering processes (i.e., Rayleigh–Mie scattering, Raman scattering, resonant scattering) that return photons into the receiver; $T_{up}^{\lambda_o}(s)$ and $T_{down}^{\lambda}(s)$ are the atmospheric transmission function along the travel forth and back from s range, and $d\Omega/4\pi$ is the solid angle subtended the receiver $(\propto 1/s^2)$.

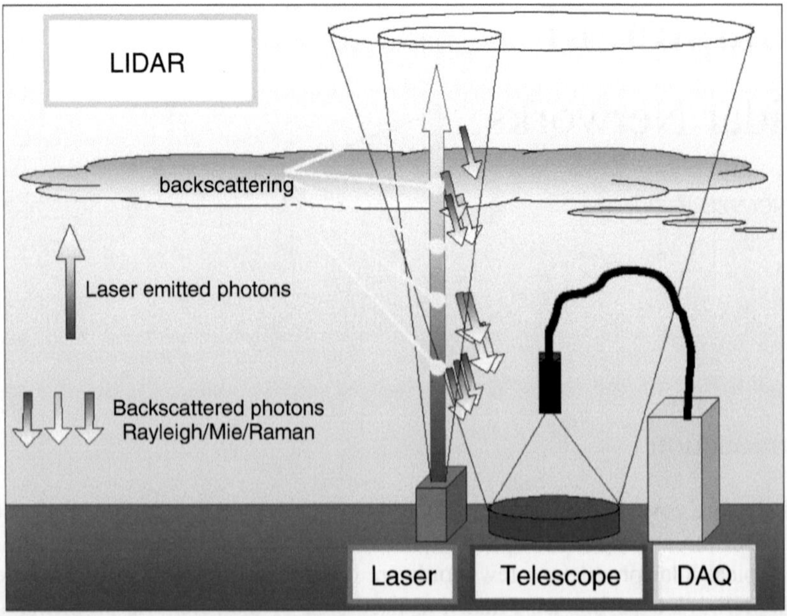

FIGURE 1. Simple representation of lidar sampling of the atmosphere, break DAQ: data acquisition system.

The baseline design of a lidar system consists of a laser, which transmits short light pulses in the atmosphere, and a receiver, usually a telescope and a detector, which collects the backscattered returns as a function of the (light pulse) time of flight (i.e., range from lidar). Quite diffuse commercial interference filters and dichroic beam splitters allow us to design a spectrometer-wise receiver: it is possible to discriminate the lidar returns according to the wavelengths.

Standard Lidar Systems

This paragraph reports a short review of the standard lidar applications devoted to detect the atmospheric composition and structure. The different atmospheric components can contribute to the backscattering and transmission terms into the lidar equation. Figure 2 illustrates various lidar applications.

Aerosol Raman Lidar

A relatively new and powerful lidar technique is based on the Raman backscattering of some molecular components of the atmosphere. This paragraph is dedicated to a simple outline of a typical aerosol Raman lidar system

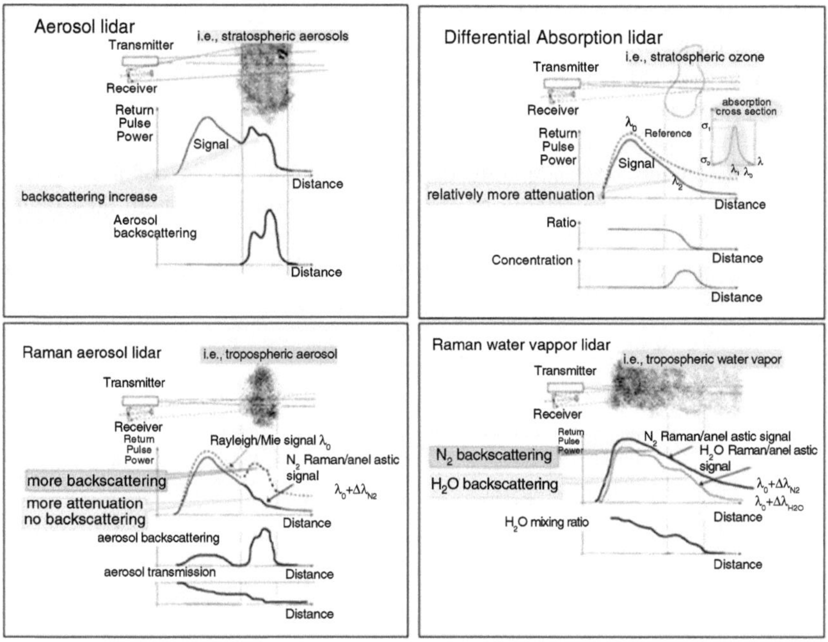

FIGURE 2. Aerosol lidar (top-left panel): an increase in the aerosol concentration reflects directly into the backscattering term of lidar return. Differential absorption lidar (top-right panel): this system makes use of two source wavelengths, one is strongly absorbed by the molecular specie to be measured (i.e., O_3), the other shows a corresponding negligible absorption; the two lidar signals differ in the transmission part, from this difference the concentration of the molecular specie can be retrieved. Raman aerosol lidar (bottom-left panel): it is one source system, but the detection of lidar returns is performed also at a different wavelength according to the Raman inelastic scattering (i.e., from N_2 or O_2 Raman scattering, more details are discussed in the next paragraphs); the Raman lidar signal is affected by the aerosol presence only in the transmission part. Raman water vapor lidar (bottom-right panel): the lidar collects signals at specific Raman shifted wavelengths (i.e., N_2 and H_2O vapor), in this case the transmission part of the lidar return have the same behavior, as a consequence, a direct comparison of the Raman backscattered signals and a calibration procedure allow us to estimate the relative fraction of H_2O.

The molecular scattering in the atmosphere consists of Rayleigh (elastic) scattering and rotovibrational Raman (anelastic) scattering. What is called Rayleigh scattering consists of rotational Raman lines and the central Cabannes line. In Raman scattering the cross section is about 3 orders of magnitude smaller than the corresponding Rayleigh cross section, and the scattered signal consists of radiation that has suffered a frequency shift that is characteristic of the stationary energy states of the irradiated molecule.

The Raman spectroscopy enables a trace constituent to be both identified and quantified relative to the major component of a mixture. As an example: the nitrogen and oxygen molecules (whose atmospheric mixing ratios are constant and known) show Raman shifts (vibrational-rotational transitions) of about 2327 cm^{-1} and 1556 cm^{-1}, respectively.

Focusing on the lidar sounding of high-resolution profiles of atmospheric aerosols; the advantages offered by the combined Raman–Rayleigh lidar detection are quite evident: more information about the measured portion of the atmosphere is available. Some details are reported to make clear this point: with the same formalism used in the previous section, the elastic/Rayleigh lidar signal has the following form:

$$L^{\lambda_o}(s) = L_o^{\lambda_o} \cdot T_{mol}^{\lambda_o}(s) \cdot T_{aer}^{\lambda_o}(s) \cdot \left[\sigma_{\text{mol}}^{\lambda_o}(\pi) \cdot n_{mol}(s) + \int_0^\infty dr \pi r^2 Q_{bck}(r, m, \lambda_o) n_{aer}(s, r) \right] \cdot \frac{d\Omega}{4\pi} \cdot c\tau_L \cdot T_{mol}^{\lambda_o}(s) \cdot T_{aer}^{\lambda_o}(s) \tag{2}$$

where

$L_o^{\lambda_o}$ number of photons emitted by the laser;

$\sigma_{mol}{}^{\lambda}{}_o(\pi)$ Rayleigh backscattering cross section:

$\sigma_{mol}^{\lambda_o}(\pi) = \frac{P(\pi)}{4\pi}\sigma_{mol}^{\lambda_o}$, $P(\pi)$ is the backscattering phase function;

$Q_{bck}(r,m,\lambda_o)$ Mie backscattering efficiency;

$T_{mol}^{\lambda}(s)$ molecular (Rayleigh + vibrational Raman) scattering transmission at wavelength λ;

$T_{aer}{}^{\lambda}(s)$ aerosol scattering transmission;

$d\Omega$ solid angle subtended by receiver;

τ_L laser pulse duration (c is the speed of light); and the atmospheric transmission functions have the following main dependence from the atmospheric components:

$$T_{mol}^{\lambda}(s) = \exp\left(-\int_0^s \sigma_{mol}^{\lambda} n_{mol}(s)\, ds \right) \tag{3}$$

$$T_{aer}^{\lambda}(s) = \exp\left(-\int_0^s \left[\int_0^\infty dr \pi r^2 Q_{ext}(r, m, \lambda) n_{aer}(s, r) \right] ds \right) \tag{4}$$

n_{mol} atmospheric molecular number density;

n_{aer} aerosol size distribution: number density of aerosol with radius between r and $r{+}dr$;

$\sigma_{mol}{}^{\lambda}$ Rayleigh total cross section;

$Q_{ext}(r,m,\lambda)$ Mie extinction efficiency of an aerosol of radius r, and refractive index m.

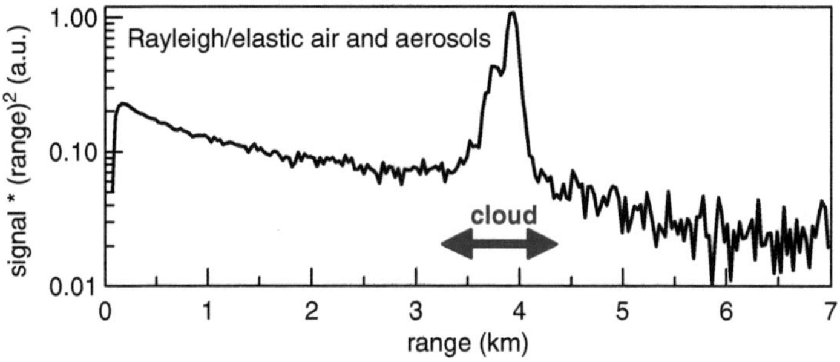

FIGURE 3. Example of elastic signal of a vertical lidar sounding the troposphere in presence of a typical low-level cloud. This profile has been collected on October 8, 2001, with the UV Raman/Rayleigh lidar at L'Aquila, Italy. The signal is the summation of about 54,000 laser shots (about 30 min); the vertical resolution is 30 m.

An example of Rayleigh lidar return is shown in Figure 3. These data have been taken in presence of a low-level cloud, within the cloud range the signal enhances due to the increasing of the effective aerosol backscattering.
The anelastic/Raman lidar signal of the ith molecule is

$$L^{\lambda_i=\lambda_o+\Delta\lambda_i}(s) = L_o^{\lambda_o}\cdot T_{mol}^{\lambda_o}(s)\cdot T_{aer}^{\lambda_o}(s)\cdot\left[\sigma_{\text{Raman}}^{\lambda_i}(\pi)\cdot n_i(s)\right]\cdot \tag{5}$$

$$\frac{d\Omega}{4\pi}\cdot c\tau_L\cdot T_{mol}^{\lambda_i}(s)\cdot T_{aer}^{\lambda_i}(s)$$

with
$\Delta\lambda_i$ is the Raman shift of the ith molecule

$$(\Delta\nu_{N_2}\approx 2327cm^{-1};\Delta\nu_{O_2}\approx 1556cm^{-1},\Delta\lambda_{N_2}=\left[\left(\frac{1}{\lambda_o}-\Delta\nu_{N_2}\right)^{-1}-\lambda_o\right];$$

$$\Delta\lambda_{O_2}=\left[\left(\frac{1}{\lambda_o}-\Delta\nu_{O_2}\right)^{-1}-\lambda_o\right]$$

$\sigma_{Raman}^{\lambda_i}(\pi)$ is the Raman backscattering cross section of the ith molecule: e.g., for N_2: $\approx 3.5\times10^{-30}$ cm^2 sr^{-1}, for O_2: $\approx 4.6\times10^{-30}$ cm^{2}sr^{-1} at $\lambda = 337$ nm.

Figure 4 shows the N_2 Raman lidar signal corresponding to the measurement of Figure 3, the cloud presence appears as an attenuation of the lidar return. Comparing Figures 3 and 4, it is evident that in the elastic channel the backscattering contribution of the cloud dominates, while in the Raman anelastic channel the main feature is the extinction in the cloud range. In general, the information retrieved by coupling Rayleigh and Raman signals are complementary and might allow a quicker estimation of the atmospheric transmissivity.

It should be noted that in the elastic channel the aerosol scattering contribution is present in both the transmission and backscattering components, while in the

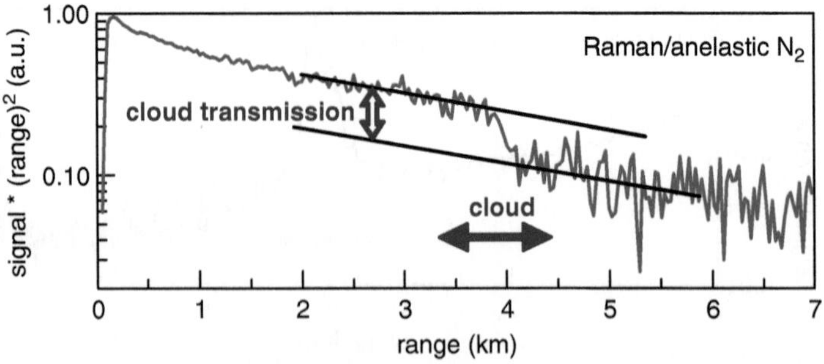

FIGURE 4. As Figure 3, but for the anelastic/Raman signal of nitrogen. Note the attenuation of the Raman signal in the cloud range.

Raman channels it appears only in the transmission. Therefore, quantitative measurements of aerosol optical properties using a lidar system that measures only aerosol elastic backscatter requires accurate system calibration and assumptions regarding aerosol optical properties (i.e., relationship between the aerosol backscatter and transmission). A combined Raman–Rayleigh lidar allows us to greatly reduce the dependence of the reconstructed transmission function on calibration and assumptions, since both aerosol transmission and backscatter are independently and directly measured.

Aerosol Characterization by Lidar

A short discussion of the performances of Raman lidar in retrieving the aerosol optical properties follows in the next sections. It is assumed that the atmospheric molecular number density is known, as well as the details of the Rayleigh scattering. To be more synthetic, we define the following quantities widely used within the lidar community:

$$\beta_{aer}^{\lambda_o}(s) = \int_0^{\infty} dr \pi r^2 Q_{bck}(r, m, \lambda_o) n_{aer}(s, r) \qquad (6)$$

$$\alpha_{aer}^{\lambda_o}(s) = \int_0^{\infty} dr \pi r^2 Q_{ext}(r, m, \lambda_o) n_{aer}(s, r) \qquad (7)$$

$\beta_{aer}^{\lambda_o}(s)$ is the aerosol volume backscatter coefficient, and $\alpha_{aer}^{\lambda_o}(s)$ is the aerosol extinction coefficient.

Rayleigh Lidar

If only the elastic signal is available, to estimate the aerosol transmission, $T_{aer}^{\lambda_o}(s)$, one must assume the so-called lidar ratio, LR in Eq. (8), which is the relationship between the aerosol extinction and backscatter:

$$LR = \frac{\alpha_{aer}^{\lambda_o}(s)}{\beta_{aer}^{\lambda_o}(s)} \qquad (8)$$

This makes it possible to solve Eq. (2) for the aerosol extinction. The obtained expression is the standard Bernoulli equation that can be integrated (Klett method), knowing a calibration height, where the boundary value of aerosol extinction can be estimated. This inversion method suffers from the fact that two physical quantities, the aerosol backscatter and the aerosol extinction, must be determined from only one measured lidar signal.

Depending on the aerosol type, the lidar ratio can vary in a wide range, and in practice, no information on the required height profile of the LR is usually available; on the other hand, if the best available constant lidar ratio is used, large uncertainties in the aerosol extinction profile remain due to the height dependence of LR.

Anelastic/Raman Lidar (Additional Detection of N_2 Raman Scattering)

In this case, the anelastic (due to molecular nitrogen Raman scattering) lidar signal is known [Eq. (5)]. To derive $T^{\lambda_o}_{aer}(s)$, in addition to the calibration value, the wavelength dependence of the aerosol extinction should be known. This is done assuming the exponent, k:

$$k = \frac{\log\left(\dfrac{\log(T^{\lambda_o}_{aer}(z))}{\log(T^{\lambda_{N2i}}_{aer}(z))}\right)}{\log\left(\dfrac{\lambda_o}{\lambda_{N_2}}\right)} \tag{9}$$

Equation (5) can be solved for $T_{aer}{}^{\lambda}{}_{o}(s)$:

$$\log\left(T^{\lambda_o}_{aer}(s)\right) = \frac{1}{1+(\lambda_o/\lambda_{N_2})^k}\log \left[\frac{L^{\lambda_{N2}}(s)}{L^{\lambda_n}_a \cdot \sigma^{\lambda_{N2}}_{\text{Raman}}(\pi)\cdot n_{N_2}(s)\cdot \dfrac{d\Omega}{4\pi}}[T^{\lambda_n}_{mol}(s)]^{\left[1+\left(\lambda_n/\lambda_{N2}\right)^{1}\right]}\right] \tag{10}$$

For aerosol particles and water droplets with diameters comparable with the measurement wavelength $k = 1$ is appropriate.

An Example with Real Lidar Signals

This example refers to a real standard session of measurements (UV Rayleigh/Raman lidar located at L'Aquila, Italy, 42.35N, 13.22E, site elevation 683m a.s.l.) during a clear-sky night. Figure 5 shows the lidar signals.

Coupling the analysis of the Rayleigh/elastic and Raman/anelastic signals, it is possible to retrieve the aerosol extinction and backscatter coefficients, and the lidar ratio: these measurements are shown in Figure 6.

A sketch of the procedure is the following (the atmosphere density is known):

- the extinction is derived from the anelastic signal assuming the scaling with wavelength ($k = 1$);

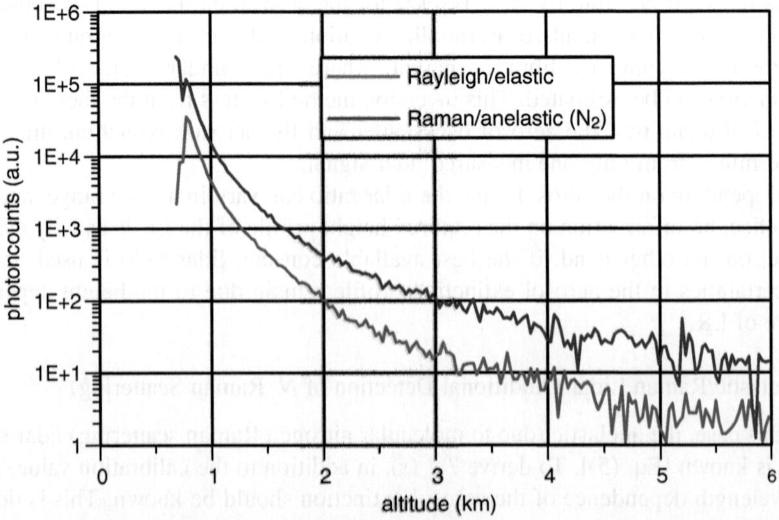

FIGURE 5. Typical elastic and anelastic lidar signals of the UV Rayleigh/Raman lidar located at L'Aquila, Italy, during a clear sky night (September 13, 2001). The vertical resolution is 30m, 54,000 laser shots are summed (about 30 min), and the background noise is subtracted. The laser source energy is about 30mJ per pulse, the emission rate is 30Hz, the signals are collected with a 20cm Ø parabolic mirror. The Rayleigh/elastic and the N_2 Raman/anelastic signals are attenuated by a factor of $\sim 5\times10^{-7}$ and $\sim 6\times10^{-4}$, respectively, due to the effective transmission of the optical components and using the appropriate interference and neutral density filters in front of the PMT detectors. Note that below ~900m (~250m range from the lidar) the overlap between receiver field of view and laser beam is incomplete.

- this extinction profile is used in the transmission part of the elastic signal and the backscattering coefficient can be determined;
- the lidar ratio is evaluated according to Eq. (8).

Lidar Networks

This paragraph is dedicated to a *partial* summary of the existing lidar networks, few details about operations and status will enlighten the advantages and the strategies of lidar networking.

NDSC, Network for Detection of Stratospheric Changes—Lidar Network

NDSC includes well-established lidar network monitoring of

- Ozone lidar (vertical profiles of ozone from the tropopause to at least 40km altitude; in some cases tropospheric ozone will also be measured)

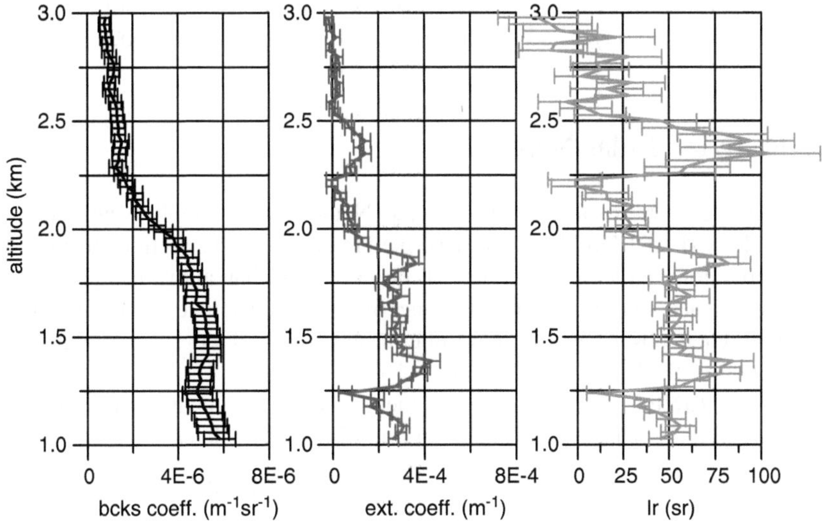

FIGURE 6. Aerosol products derived from the signals in Figure 5 combining the inversion of anelastic and elastic signals. From left to right: the aerosol volume backscattering coefficient, the aerosol volume extinction coefficient, and the lidar ratio versus altitude. The error bars indicate 1σ standard deviations.

- Temperature lidar (vertical profiles of temperature from about 30 to 80 km)
- Aerosol lidar (vertical profiles of aerosol optical depth in the lower stratosphere)
- Water vapor lidar (vertical profiles of water vapor in the lower stratosphere)

The NDSC sites are widely distributed and about one-fourth of them has an operative lidar system; more information can be found at http://www.ndsc.ncep.noaa.gov/.

AD-Net, Asian Dust Network—Lidar Network Observation of Kosa in Japan

It is constituted by a coordinated network of Rayleigh lidars, distributed among Malta, China, South Korea, Japan, and USA, its main purposes concern:

- the mechanism of transportation of Asian dust
- evolution of Asian dust aerosols during the transportation
- impact of Asian dust to ocean Bio-products
- dust-clouds (water clouds/cirrus) interaction

http://info.nies.go.jp:8094/kosapub/.

The Micropulse Lidar Network—MPLNET

This lidar network is constitutes of standard and high sophisticated *stand-alone* Rayleigh lidars. The lidars'distribution and performances make the network well suited for intensive campaigns for cloud and aerosol structure studies into the planetary boundary layer and lower troposphere, all around the globe.

EARLINET Lidar Network

The European Aerosol Research Lidar Network to Establish an Aerosol Climatology (EARLINET) will be described in details in this section, also giving the motivations and the strategies of observations that allow us to perform specific science studies. Highlights concerning the advantages of using a lidar network when observing the atmospheric aerosols will be clearer within the description of the EARLINET results.

The EARLINET project is based on the following statements (extracted from the EARLINET documents, available at http://lidarb.dkrz.de/earlinet/):

- a good way to obtain information on aerosols with high spatial resolution is the lidar technique;
- the Raman lidar technique allows us to determine physical and optical properties of aerosols;
- a coordinated ground-based network has extra benefits;
- the EARLINET studies include, among others, the establishing of local aerosol climatology, the observation of long-range transport of Saharan dust layers, and the modification of aerosol properties during their passage over Europe;
- lidar measurements may give a significant gain of knowledge in aerosol science;
- the progress will be even stronger if lidar and passive remote sensing techniques are used complementary and are applied in a synergistic way;
- a quantitative data set describing the aerosol vertical, horizontal, and temporal distribution, including its variability on a continental scale, is necessary. Such a data set could be used to validate and improve models.

The scientific objectives and approaches of EARLINET are as follows:

- to provide aerosol data with unbiased sampling, together with comprehensive analyses of these data;
- the objectives will be reached by implementing a network of more than 20 lidar stations distributed over most of Europe;
- special care will be taken to assure data quality, including intercomparisons at instrument and evaluation levels;
- a major part of the measurements will be performed according to a fixed schedule to provide an unbiased statistically significant data set;
- additional measurements will be performed to specifically address important processes that are localized in either space or time.

The project has given an important contribution to the improved model treatment of physical processes, in particular aerosols in the boundary layer. The collected data can also be used to improve the quality of a number of satellite retrieval systems that are affected by the presence of aerosols. In addition, cooperation within the network has led to a very efficient transfer of know-how in two important areas: advanced remote sensing using high-tech instruments, and the application of these techniques to address complex environmental problems.

EARLINET Data Quality: Instruments

To achieve a homogeneous data set, and to make sure that all systems work well, a number of intercomparison experiments were performed that tested at least two systems at the same time in the same place. The different lidar groups compared their systems with quality-assured lidar systems. In most cases the predefined quality criteria could be met from the beginning. The aerosol backscatter measurements showed deviations of less than 10% in most cases. The standard deviations were typically below 25%. System precision, including algorithms, could be estimated to be better than 20% in many cases and to even better than 10% in the planetary boundary layer.

EARLINET Data Quality: Algorithms

Besides instrument intercomparison, a basic exercise to ensure the quality of network measurements is the comparison of the algorithms that are used to calculate the optical parameters from lidar signals. For the aerosol backscatter coefficient retrieved from the Rayleigh channel of the lidar systems, the mean difference between the exact solution and the estimations of various lidar groups can be regarded as negligible when they are compared to the uncertainties caused by bad estimation of the input parameters like lidar ratio and reference value. The intercomparison of the aerosol extinction evaluation starting from nitrogen Raman lidar signals and of the retrieval of the aerosol backscatter by using the combined Raman–Rayleigh backscatter lidar technique shows that the aerosol extinction evaluations can be accomplished with good accuracy for most EARLINET groups. The retrieved aerosol extinction profiles differ from the solution within 15% and 20% in the 400–4000m height range. This intercomparison shows satisfactory results also for the aerosol backscatter coefficient, in particular, also without any reference value for the backscatter, the retrieval of the aerosol backscatter starting from simultaneous Raman–Rayleigh lidar signals is satisfactory, demonstrating how much more powerful the Raman elastic-backscatter lidar technique is compared to the case in which only Rayleigh lidar signals are available.

EARLINET Results

The main results of EARLINET are reported in the next sections: along the discussion of the local (i.e., at L'Aquila lidar site) PBL aerosol climatology, of some episodes of Saharan dust transportation and of the aerosol distribution on EARLINET (i.e., continental) scale, it is made evident the advantages of co-ordinated observations.

PBL Aerosol Climatology at L'Aquila Lidar Site

Each EARLINET lidar station takes measurements at least two times a week. When only the clear sky soundings are considered, the time series of relevant quantities show seasonal cycles (Figure 7), but also statistical analysis of the different data can be performed (e.g., aerosol backscatter coefficient, aerosol extinction, aerosol optical depth, dust layer height, or *lidar-oriented* quantities like LR, which is connected to the Mie scattering phase function).
Looking at Table 1, the main results are, for example:

- high standard deviations, that means short time (day by day) variability;
- high skewness of the data frequency distribution, that suggests that the data are log-normal distributed and this is probably peculiar of geophysical data.

Example of Saharan Dust Transportation Monitored by EARLINET

Figure 8 reports an example of a case study for a Saharan dust outbreak over Europe, the coordinated lidar network made possible to follow the evolution of the Saharan dust distribution on time and geographical extent.

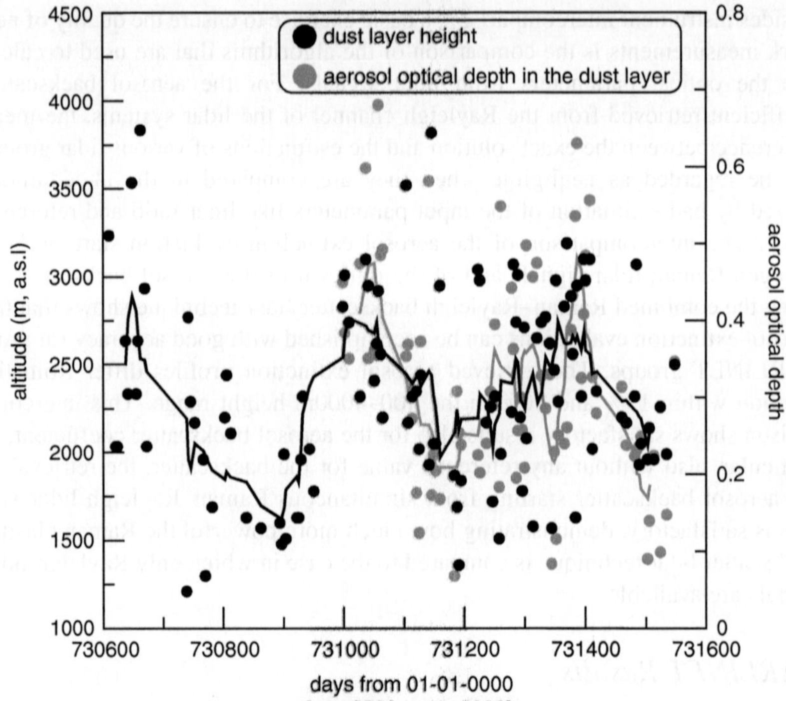

FIGURE 7. Dust layer heights (DLH, the height of lowest layer that generally contains most of the aerosol) and aerosol optical depth in DLH as a function of time (May 2000–Nov. 2003). Solid lines: running average, window width 7 points.

TABLE 1. A Resume of the Statistical Analysis of EARLINET Lidar Observations at L'Aquila

Dust layer STATISTIC	Backscatter (10^6 m^{-1} sr^{-1})	Extinction (10^4 m^{-1})	Lidar Ratio (sr)
Mean	3.39/3.67	1.96	54
Standard deviation	1.85/1.85	1.04	9
Median	3.03/3.31	1.77	53
Skewness	1.06/1.15	1.4	0.32
# Cases	108/75	75	75

Dust layer STATISTIC	Aerosol optical depth	Dust layer height (m)
Mean	0.32	2340
Standard deviation	0.14	555
Median	0.30	2288
Skewness	0.49	0.35
# Cases	75	108

The appearance of the layers can be analyzed also using air mass trajectory reconstruction. The analysis of the data collected during the 3 years of operation of EARLINET made possible, for the first time, the simultaneous estimation of the horizontal and vertical extent of free tropospheric Saharan dust layers over Europe, as well as to provide important input for radiative transfer models and atmospheric chemistry transport models over Europe.

The maximum intensities of the Saharan dust outbreaks are in S-SE europe in summer, autumn, and spring periods, in summer and spring for SW europe.

PBL Aerosol Climatology on Continental Scale

A statistical analysis concerning the vertical distribution of the volume light-extinction coefficients of particles derived from Raman lidar measurements at 10 EARLINET stations (Table 2) is presented here.

The profiles were measured on a fixed schedule with up to two measurements per week; they typically covered the height range from 500m to 6000m above ground level (agl). The analysis is made for the planetary boundary layer (PBL) as well as for several fixed layers above ground.

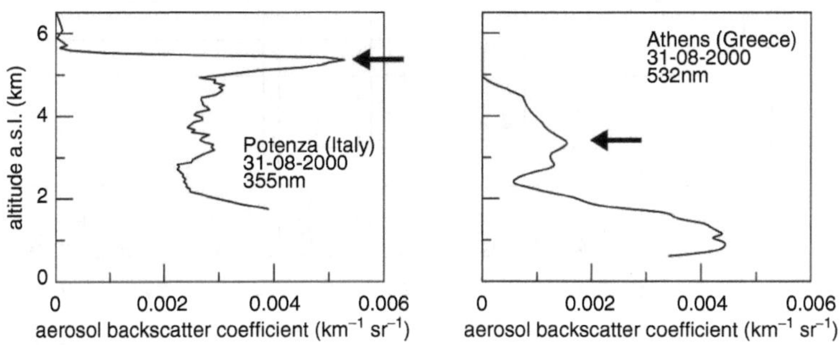

FIGURE 8. Saharan dust layers observed simultaneously over Potenza, Italy (left) and over Athens, Greece, at 19:00UT on August 31, 2000.

TABLE 2. Number of Measurements, Covered Time Periods, and Number of Considered Measurements in Three Altitude Intervals for 10 EARLINET Raman Lidar Stations

Station	acronym	covered time	all	summer	winter	0–1 km	1–2 km	2–5 km
Aberystwyth	ab	5/00–11/02	55	34	21	53	55	27
Athens	at	11/00–11/02	81	45	36	79	61	9
Hamburg	hh	5/00–11/02	109	71	38	103	84	48
Kühlungsborn	kb	5/00–11/02	62	39	23	0	45	50
L'Aquila	la	5/01–11/02	75	41	34	75	75	7
Lecce	lc	5/00–8/02	166	94	72	162	137	24
Leipzig	le	5/00–12/02	83	55	28	8	77	81
Naples	na	10/00–12/02	145	64	81	141	143	69
Potenza	po	5/00–12/02	88	60	28	26	85	75
Thessaloniki	th	02/01–11/02	58	31	27	50	55	30

The results show typical values of the aerosol extinction coefficient and the aerosol optical thickness (Figure 9 and Figure 10) in different parts of Europe, with highest values in SE Europe and lowest values in the NW part. It is found that higher aerosol optical thickness in southern Europe compared to the northern part is mainly attributed to larger amounts of aerosol in higher altitudes. At 9 of the 10 sites the frequency distribution of the aerosol optical thickness in the planetary boundary layer follows a lognormal distribution at the 95% significance level.

The aerosol optical thickness in the PBL shows an annual cycle with higher values in summer than in winter for most of the stations. Highest values can usually be found in late summer (August/September).

Regular aerosol lidar measurements give the opportunity to provide a statistics on the PBL height, which can be determined from the range corrected lidar signal.

The PBL height shows in most cases a clear annual cycle with higher values in summer than in winter (Table 3). The variability of the PBL height is high.

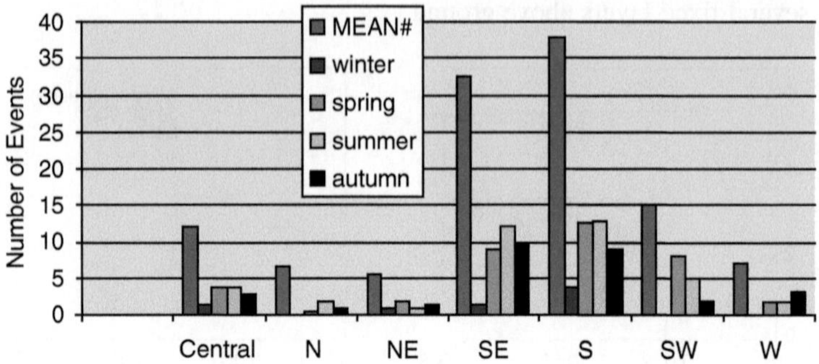

FIGURE 9. Seasonal variability of Saharan dust outbreaks observed within EARLINET (2000-2003) (normalized number of events).

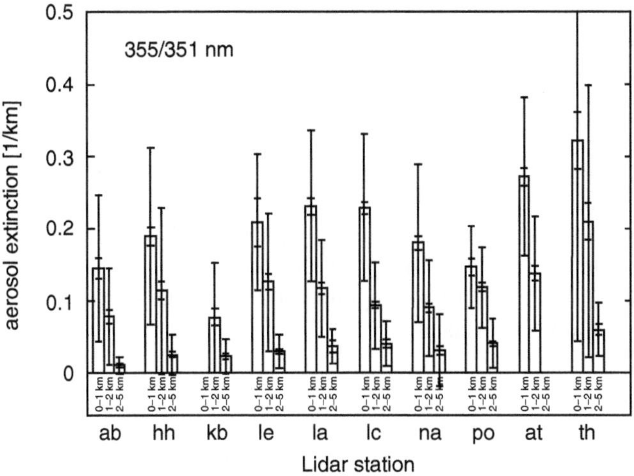

FIGURE 10. Mean aerosol extinction coefficient in three layers, 0–1 km, 1–2 km, and 2–5 km above ground level. Thin (outer) error bars denote the standard deviation of the individual values, representing the variability between the profiles. Thick (inner) error bars give the standard deviation of the mean.

A Few Conclusions

Lidar network systems offers powerful strategies and procedures within atmospheric science studies. Although a single lidar system locally measures an atmospheric component (but at high resolution in time and vertical extent), in a lidar network the coordination of observing periods coupled, for example, with air mass trajectory reconstruction, increase the density of geographical coverage. From this side, the activities of a lidar network could be helpful to validate and improve models, as well as, to calibrate and validate satellite observations.

TABLE 3. Statistical Parameters of the PBL Height for 10 EARLINET Stations

Station	PBL-height/m agl	Std. Dev/m	Rel. Std. Dev	Skewness
Aberystwyth	1204	481	0.40	0.62
Humburg	1242	506	0.41	0.62
Kühlungsborn	1984	566	0.29	0.36
Leipzig	1937	796	0.41	-0.07
L'Aquila	1742	509	0.29	0.18
Lecce	1261	599	0.48	1.29
Naples	1442	626	0.43	0.70
Potenza	1542	304	0.20	0.69
Athens	1172	319	0.27	1.16
Thessaloniki	1363	416	0.31	0.74

The discussion of the operations and results of an European Raman lidar network for the aerosol monitoring (EARLINET) has enlightened the features (from local to continental scale) of the science studies that could be performed.

Acknowledgements Most of the description of EARLINET objectives, operations, and results have been taken from the EARLINET official documents and also from some of the project scientific papers. The EARLINET people and its coordinator (Jens Bosenberg) are gratefully acknowledged. I am also grateful to Guido Visconti for his continuos support and to Piero Di Carlo for his never ending and contagious enthusiasm.

Essential References

A monograph on laser remote sensing and its applications. It reviews the basic physics needed to understand the subject:

Laser Remote Sensing: Fundamentals and Applications, Raymond M. Measures, New York, Wiley, 1984; reissued, Malabar, Fla., Krieger Pub. Co., 510 pages, 1992.

Papers reporting the different (Lidar Network) activities within EARLINET:

Vertical aerosol distribution over Europe: Statistical analysis of Raman lidar data from 10 European Aerosol Research Lidar Network (EARLINET) stations, Matthias V., Balis D., Bosenberg J., Eixmann R., Iarlori M., Komguem L., Mattis I., Papayannis A., Pappalardo G., Perrone M.R., Wang X., *Journal of geophysical research atmospheres*, **109** (D18), D18201, 2004.

Aerosol lidar intercomparison in the framework of the EARLINET project. 1. Instruments, Matthias V., Freudenthaler V., Amodeo A., Balin I., Balis D., Bosenberg J., Chaikovsky A., Chourdakis G., Comeron A., Delaval A., De Tomasi F., Eixmann R., Hagard A., Komguem L., Kreipl S., Matthey R., Rizi V., Rodrigues J.A., Wandinger U., Wang X., *Applied Optics*, **43** (12), 2578–2579, 2004.

Aerosol lidar intercomparison in the framework of the EARLINET project. 2. Aerosol backscatter algorithms, Bockmann C., Wandinger U., Ansmann A., Bosenberg J., Amiridis V., Boselli A., Delaval A., De Tomasi F., Frioud M., Grigorov I.V., Hagard A., Horvat M., Iarlori M., Komguem L., Kreipl S., Larchevque G., Matthias V., Papayannis A., Pappalardo G., Rocadenbosch F., Rodrigues J.A., Schneider J., Shcherbakov V., Wiegner M., *Applied Optics*, **43** (4), 977–989, 2004.

Aerosol lidar intercomparison in the framework of the EARLINET project. 3. Raman lidar algorithm for aerosol extinction, backscatter, and lidar ratio, Pappalardo G., Amodeo A., Pandolfi M., Wandinger U., Ansmann A., Bosenberg J., Matthias V., Amirdis V., De Tomasi F., Frioud M., Iarlori M., Komguem L., Papayannis A., Rocadenbosch F., Wang X., *Applied Optics* **43**(28), 5370–5385, 2004.

Chapter 12

U.S. Federal and State Monitoring Networks

KENNETH L. DEMERJIAN

Introduction

The identification and measurement of chemical constituents in the atmosphere are essential in defining and understanding the state of environmental air quality and its changes with time. Environmental monitoring networks generally fall within two broad classes, those which are designed to determine the physical and chemical state of the environment (e.g., air quality, meteorology, water quality, etc.) and those which are designed to determine the ecological state of the environment (species diversity, soil erosion, biomass productivity, etc.). This discussion focuses on national air-quality monitoring networks in the United States which have been specifically designed to address air-quality issues associated with criteria pollutants or pollutants with suspected health impacts as identified under the U.S. Clean Air Act and its amendments.

U.S. Federal and State Monitoring Networks

Air-quality monitoring networks in the U.S. are designed to address multiple objectives. These include (1) to provide a national database for determining air quality in major metropolitan areas, (2) to observe pollution trends in urban and nonurban areas, (3) to assess compliance or progress made toward meeting air-quality objectives/ standards, (4) to provide data for implementation of emergency control plans in the prevention or mitigation of air pollution episodes, and (5) to provide data bases for determining the chemical loading of ecosystems through atmospheric deposition and its trend [Demerjian, 2000]. The broad monitoring objectives have resulted in the development of independent networks, which might address one or two objectives, but which has been criticized for not being integrated more effectively. That being said the networks historical emphases have predominantly been used to measure air quality to determine (1) maximum concentrations within specified spatial domains, (2) representative concentrations for exposure assessments, (3) source—receptor relationships mainly pertaining to the impact of specific sources on local air quality, (4) background concentrations and temporal trends. The backbone of the air-quality monitoring system is made up of two major networks, the State and Local Air Monitoring Stations (SLAMS) and the National Air Monitoring Stations (NAMS). The former network is operated by state and local agencies drawing from U.S. EPA federal assistance grants and the latter, NAMS, also operated by state agencies, but overseen and used by the U.S. EPA for trend analysis and tracking air quality across the nation. The SLAMS, initiated in 1980, now consists of ~4000 monitoring stations operated by state and local air pollution control agencies. The NAMS, a subset of the SLAMS network, consists of 1080 stations located in urban and multi-source influence areas. The principal purpose of SLAMS/ NAMS is to provide a nationwide database for determining air quality in major urban centers and in regional/ rural environs of the United States and provide data in support of state implementation plans (SIP). Monitoring sites are classified based on spatial scale of representativeness (micro, middle, neighborhood, urban, regional, and national-global). Measurements include continuous O_3, NO_x, CO, and SO_2 measurements (reported hourly), PM10 and high-volume total suspended particulate matter (TSP) 24-hour integrated sample collected at least once every sixth day. All parameters are not measured at every site. Figure 1 shows SLAMS/ NAMS monitoring sites across the U.S. by parameter.

Stations (NAMS) Sites by Chemical Parameter

In the 1990 Clean Air Act Amendments [Section 182 (c)(1)], the U.S. EPA was mandated to improve monitoring of ozone and its precursors within specified

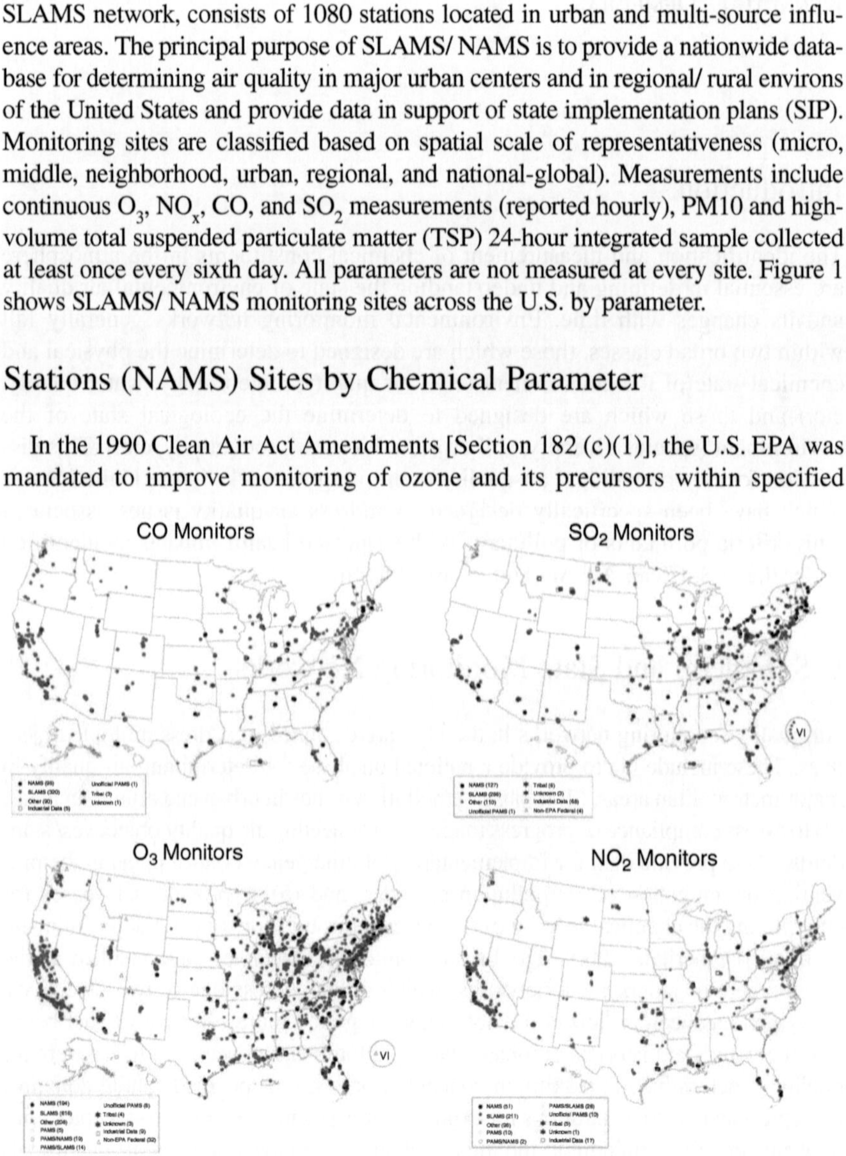

FIGURE 1. State and local air monitoring stations (SLAMS) and the national air monitoring.

ozone nonattainment areas. This action was in part a response to a pending report by the National Academy of Sciences (NRC, 1991) which indicated a significant need for improved precursor monitoring and data analyses to better address ozone air-quality management issues. The U.S. EPA initiated the Photochemical Assessment Monitoring Stations (PAMS) program in February 1993 to be implemented by state agencies starting in 1994 [U.S. EPA, 1993]. The PAMS network, designed specifically to track and assess progress in photochemical oxidant air quality measures O_3, NO, NO_x, 56 targeted hydrocarbon compounds and three carbonyl species (formaldehyde, acetaldehyde and acetone), as well as a standard set of surface meteorological measurements. PAMS, established in 1994, was to consider up to 5 monitoring stations per O_3 nonattainment area, but in practice has not exceeded three stations per nonattainment area. There are up to four classes of monitoring sites (Types 1–4), which are designed to meet unique data objectives of the network. The PAMS are operated by state and local governments, and consists of 65 sites deployed as shown in Figure 2 [U.S. EPA, 1993 & 1997]. PAMS measurements operate from May 15–September 15 providing continuous O_3, NO/ NO_x (reported hourly), 56 target hydrocarbons (hourly or 3-hour average), formaldehyde, acetaldehyde, and acetone (3-hour average), and surface meteorological parameters (hourly).

The Clean Air Status and Trends Network (CASTNet), established in 1990 under mandate of the 1990 Clean Air Act Amendments, incorporated the National Dry Deposition Network (NDDN), which was established in 1986 as part of the National Acid Precipitation Program (NAPAP). NDDN operated 6 sites in 1987. By 1991, the CASTNet program had 50 sites (see Figure 3). The purpose of the network is to provide a nationwide database of rural sites to evaluate patterns and trends in the concentration of chemical species and dry and wet chemical deposition on regional scales. Measurement parameters include precipitation chemistry and cloud water samples analyzed for pH, $SO_4^{=}$, NO_3^-, Cl^-,

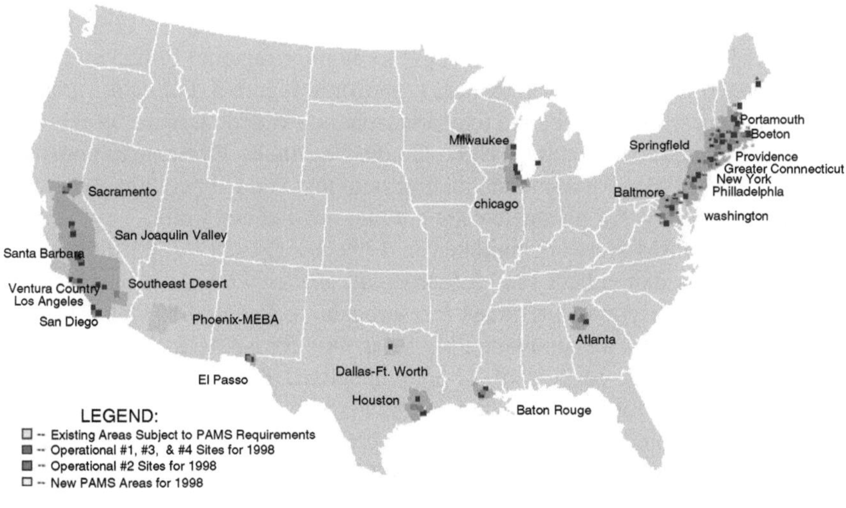

FIGURE 2. Photochemical assessment monitoring stations by operational type.

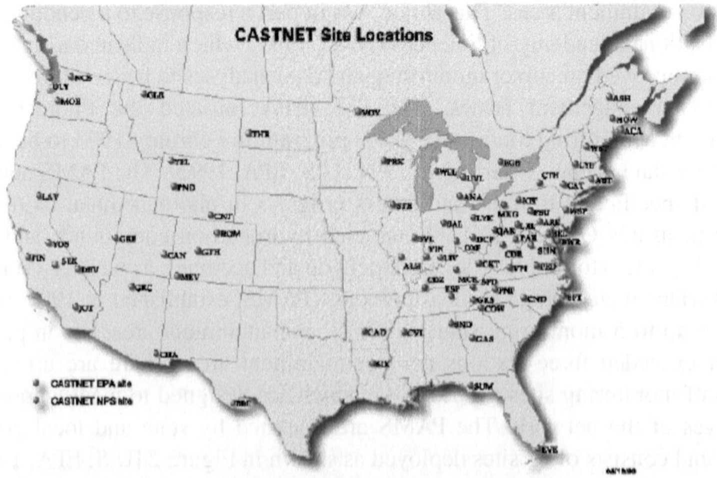

FIGURE 3. Clean Air Status and Trends Network (CASTNet).

NH_4^+, Ca^{2+}, Mg^{2+}, and K^+; Weekly particulate $SO_4^=$, NO_3^- and NH_4^+, and gaseous HNO_3 and SO_2; Continuous O_3, meteorological parameters, solar radiation. Sites provide inferential estimates of dry deposition [U.S. EPA, 1995].
Interagency Monitoring of Protected Visual Environments (IMPROVE) network, initiated in 1987, consists of 53 monitoring sites that collect both PM2.5 and PM10 from March 1988 to provide a nationwide database that supports visibility protection regulations for Federal Class I areas including the documentation of present visibility levels, identification of sources of existing manmade impairment, and track progress in achieving no manmade impairment in protected areas. Figure 4 shows Class I areas in the U.S. and the initial IMPROVE deployment. The network was expanded to 90 sites in 1999 and to 170 sites by 2001 [Malm et al., 1994; Eldred et al., 1990 and 1997] to monitor regional PM2.5 and haze. The expanded network site deployment is shown in Figure 5. Pollutant variables include hourly scattered light and total atmospheric extinction, twice weekly elemental and organic carbon, $SO_4^=$, NO_3^-, Cl^-, H, Na-Mn, Fe-Pb elements and mass for PM 2.5 and PM10., five-minute relative humidity and temperature, 3 pictures per day of visibility. All parameters are not measured at every site.

The PM2.5 FRM Network, initiated in 1999, consists of ~1200 filter based monitoring sites that collect PM2.5 (particulate matter with mean diameters of 2.5 μm or less) for gravimetric mass measurement at a centralized lab. The network, which provides a nationwide PM2.5 database for NAAQS compliance and exposure assessment, consists of 24-hour integrated PM2.5 samples collected once in three days for most samplers, with everyday samples operated at select trends sites and once-in-six-day samplers operating at select SLAMS sites for enhanced spatial representativeness. The PM2.5 FRM network site deployment as of 1999, shown in Figure 6, also provides the 3-year annual mean PM2.5 mass concentrations with respect to the PM2.5 NAAQS (i.e., 15μg/ m^3).

FIGURE 4. Interagency Monitoring of Protected Visual Environments (IMPROVE) Network Initial Deployment and U.S. Class I Areas.

In addition, in support of the revised NAAQS for PM2.5, the U.S. EPA initiated in 2000 a PM2.5 Speciation Network to consist of ~270 filter-based monitoring sites that collect PM2.5 for chemical analysis at centralized laboratories. Fifty-four of these sites constitute the speciation trends network. The primary purpose for the network is to provide a speciation database to support PM2.5

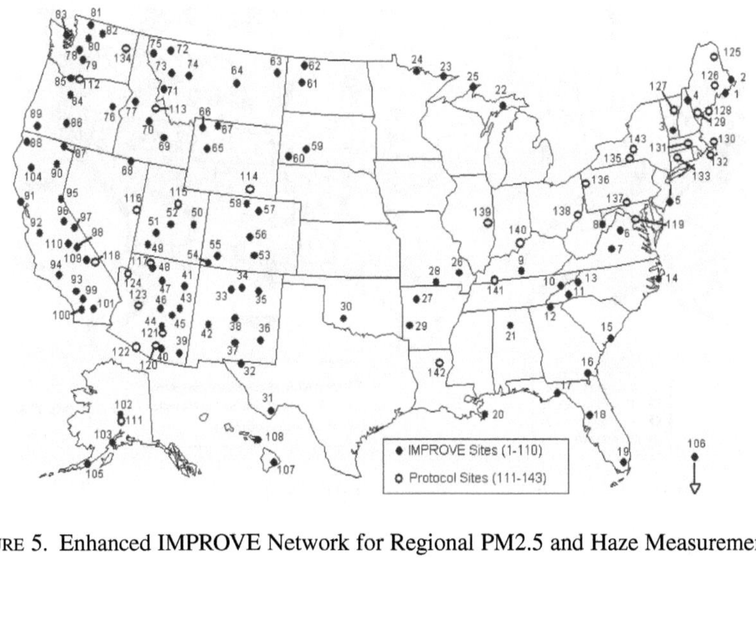

FIGURE 5. Enhanced IMPROVE Network for Regional PM2.5 and Haze Measurements.

source attribution, model, and SIP development and evaluation, exposure assessment, and health effects research. The measurement parameters include 24-hour integrated PM2.5 samples collected once in three days for most samplers, with some once-in-six-day samplers operating at select SLAMS sites for enhanced spatial representativeness. PM analysis include metals, anions and cations as described for IMPROVE PM2.5 enhanced measurements. Figure 7 shows the full deployment of PM2.5 speciation monitoring sites across the U.S. including SLAMS, STN (speciation trend network), and IMPROVE regional sites.

The last network to be discussed, the Air Toxics Monitoring Network (ATMN), is currently under development and is undergoing prototype testing at selected locations in the U.S. The proposed network and current pilot sites are shown in Figure 8.

Although there are 188 chemical species identified as hazardous pollutants, 33 have been identified as having the greatest threat to public health in the largest number of urban areas air toxics. The 33 air toxic substances are listed in table 1. The compounds to be monitored in the pilot network include a significant number of VOC, SVOC, and metals species. The most prevalent compounds observed in the 2003 pilot monitoring studies included:1,3-butadiene, benzene, xylene (*o*, *m*-, *p*-), bromomethane, carbon tetrachloride, *p*-dichlorobenzene, tetrachloroethylene, acetonitrile, acrylonitrile, acetaldehyde, and formaldehyde [U.S. EPA,2004c].

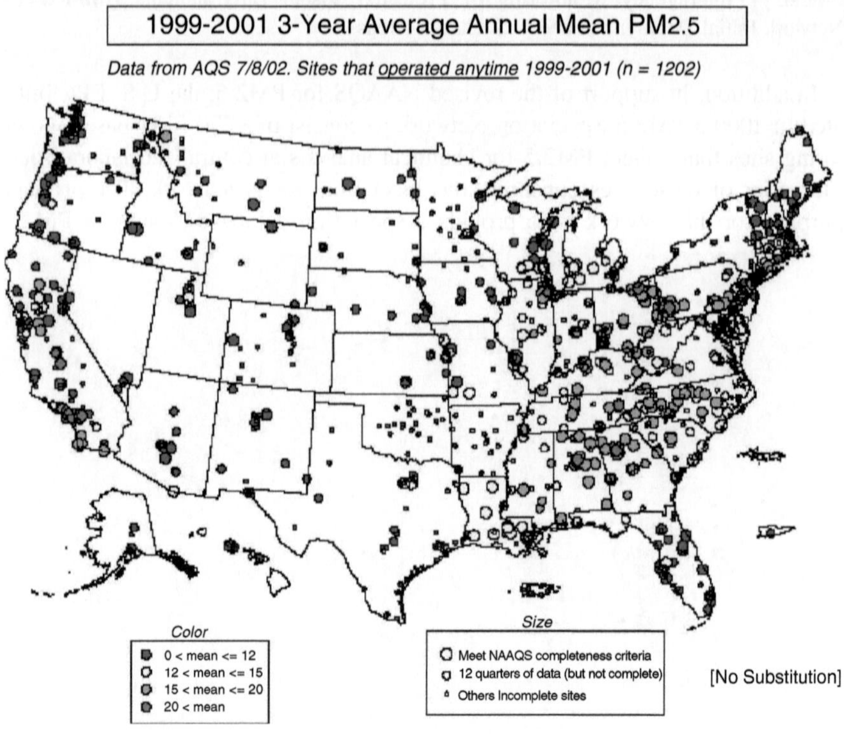

FIGURE 6. PM2.5 Federal Reference Method (FRM) Network.

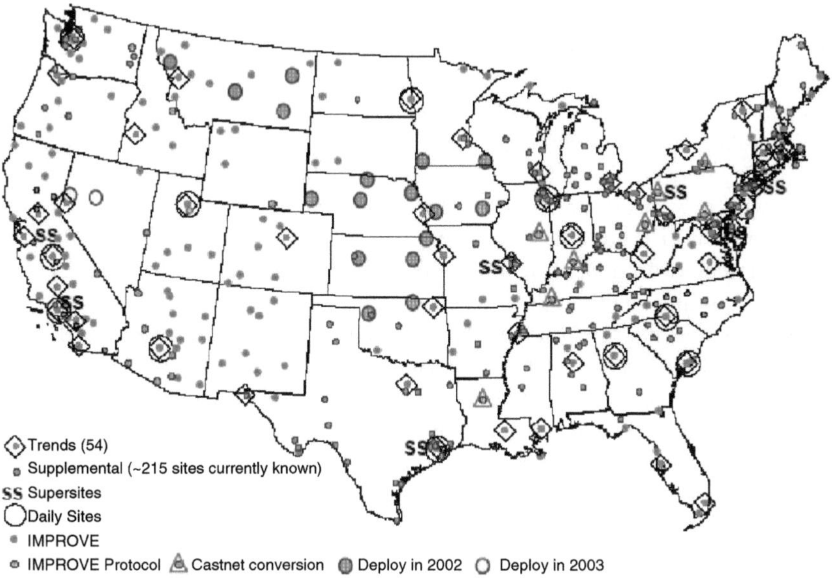

FIGURE 7. PM2.5 Speciation Monitoring Network Sites (SLAMS, Trends, and IMPROVE).

Future Monitoring Networks

The U.S. EPA is currently in the process of restructuring his monitoring activities as a result of changes in air-quality and national needs. As a result of successful air-quality management practices, most criteria pollutant measurements read well

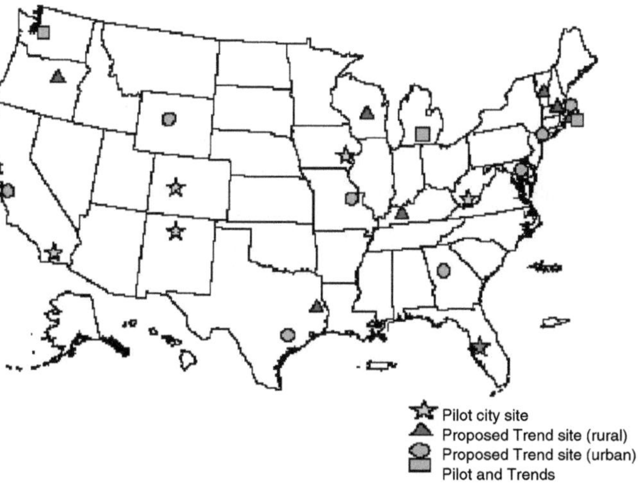

FIGURE 8. Proposed air toxics monitoring network.

below the national standards. This result suggests that a restructuring of existing monitoring activities to meet new directions and demands for air-quality measurements can be achieved without compromising requirements for current NAAQS compliance monitoring. With this in mind, the U.S. EPA has developed a National Ambient Air Monitoring Strategic Plan [U.S. EPA, 2004a] for the restructuring of the existing networks to meet new air-quality challenges as well as upgrade networks with new measurement technologies and provide more optimally located, integrated multi-pollutant monitoring sites. The proposed National Core Monitoring Network (NCore) would replace/restructure current SLAMS/ NAMS monitoring stations with a three tier monitoring system of increasing sophistication and complexity. The proposed NCore levels 1–3 monitoring sites are defined as follows. Level 1 sites, the most sophisticated and smallest in number, 6–8 sites deployed across the country, are research/technology transfer-oriented platforms accommodating extensive instrumentation and advanced measurement technologies. These monitoring sites will be used to evaluate existing and emerging measurement technologies and to provide detailed, high time-resolved measurements of the PM/co-pollutant complex and associated precursors, including detailed characterization of PM composition as a function of particle size. The specific focus of the level-1 sites will vary with location and over time, but will always be hypothesis driven and address both measurement technology and process science issues. Level 1 monitoring sites are conceptually modeled around the U.S. EPA PM2.5 Supersite programs and draw upon experiences of those operations. Level 2 sites constitute the backbone of the NCore network with the deployment of approximately 75 nationwide multi-pollutant sites, encompassing both urban (about 55 sites) and rural (about 20 sites) locations. Level 2 sites would provide an extensive set of measurements (see Table 2 for proposed parameters; EPA, 2004a) rarely found currently at any individual monitoring site.

TABLE 1. Air Toxic Pollutants (33) Identified in EPA National Assessment

1. acetaldehyde	18. formaldehyde
2. acrolein	19. hexachlorobenzene
3. acrylonitrile	20. hydrazine
4. arsenic compounds	21. lead compounds
5. benzene	22. manganese compounds
6. beryllium compounds	23. mercury compounds
7. 1, 3-butadiene	24. methylene chloride
8. cadmium compounds	25. nickel compounds
9. carbon tetrachloride	26. perchloroethylene
10. chloroform	27. polychlorinated biphenyls (PCBs)
11. chromium compounds	28. polycyclic organic matter (POM)*
12. coke oven emissions	29. propylene dichloride
13. 1, 3-dichloropropene	30. quinoline
14. diesel particulate matter	31. 1, 1, 2, 2-tetrachloroethane
15. ethylene dibromide	32. trichloroethylene
16. ethylene dichloride	33. vinyl chloride
17. ethylene oxide	

* also represented as 7-PAH

TABLE 2. NCore Level 2 Core Parameter List

Measurements	Comments
PM2.5 speciation	OC/ EC fractions, major ions and trace metals (24 hour average; every 3rd day)
PM2.5 FRM mass	typically 24 hr. average every 3rd day
continuous PM(10-2.5) mass	1 hour reporting interval for all cont. species
ozone (O_3)	all gases through cont. monitors (except HNO_3 and NH_3)
carbon monoxide (CO)	capable of trace levels (low ppb and below) where needed
sulfur dioxide (SO_2)	capable of trace levels (low ppb and below) where needed
nitrogen oxide (NO)	capable of trace levels (low ppb and below) where needed
total reactive nitrogen (NO_y)	capable of trace levels (low ppb and below) where needed
ammonia (NH_3)	through denuders; 12 samples per year @monthly average
nitric acid (HNO_3)	through denuders; 12 samples per year @monthly average
surface meteorology	wind speed and direction, temperature, pressure, RH

Level 3 sites will consist of approximately 1000 monitoring sites with attributes similar to current SLAMS deployments and a specific focus on pollutants concerned with NAAQS compliance and SIP preparations.

Summary

Extensive monitoring networks have evolved with the promulgation air-quality legislation to protect the environment in the United States. The development of National Ambient Air Quality Standards has focused networks to address these specific air pollutants somewhat independently, when in reality many of the pollutants of current interest (PM2.5, photochemical oxidants) are known to be closely coupled through their precursors and transformation processes. It is now recognized that the integration of monitoring networks to measure the critical atmospheric components associated with PM2.5, photochemical oxidants, and air toxics is most cost-effective and scientifically creditable approach. The U.S. EPA's proposed National Air Quality Monitoring Strategy moves significantly in this direction, restructuring and enhancing a significant number of existing pollutant specific monitoring sites to establish a core set of integrated multipollutant monitoring stations. The implementation the U.S. EPA NCore strategy is expected to be fully implemented by 2007.

References

Demerjian, K.L. (2000). A review of national monitoring networks in North America, Atmospheric Environment, **34**, 1861–1884.

Eldred, R.A., T.A, Cahill, R.G. Flocchini, Composition of PM2.5 and PM10 aerosols in the IMPROVE network, *J. Air & Waste Manage. Assoc.*, **47**, 194–203, 1997.

Eldred, R.A., T.A. Cahill, L.K. Wilkinson, P.J. Feeney, J.C. Chow, W.C. Malm, Measurement of fine particles and their chemical composition in the IMPROVE/ NPS networks, In Visibility and Fine Particles; Mathai, C.V., (ed.), *Air & Waste Manage. Assoc.*: Pittsburg, PA, pp.187–196, 1990.

Malm, W.C., J.F. Sisler, D. Huffman, R.A. Eldred, T.A. Cahill, Spatial and seasonal trends in particle concentration and optical extinction in the United States, *J. Geophys. Res.* **99**, 1347–1370, 1994.

National Academy of Sciences, Rethinking the Ozone Problem in Urban and Regional Air Pollution, National Academy Press, Washington, DC, 1991.

NARSTO (2000). An Assessment of Tropospheric Ozone Pollution – A North American Perspective, NARSTO Synthesis Team, July 2000. http://www.cgenv.com/Narsto/.

NARSTO (2003). Particulate Matter Science for Policy Makers—A NARSTO Assessment, February 2003. http://www.cgenv.com/Narsto/.

U.S. Environmental Protection Agency (1998a).Guidance for using continuous monitors in PM2.5 monitoring networks, OAQPS, Research Triangle Park, NC 27711, EPA-454/ R-98-012, May 29, 1998. http://www.epa.gov/ttn/amtic/files/ambient/pm25/r-98-012.pdf.

U.S. Environmental Protection Agency (1998b). Technical Assistance Document (TAD) for Sampling and Analysis of Ozone Precursors; EPA/600-R-98/161, OAQPS, Research Triangle Park, NC 27711, September 30, 1998. http://www.epa.gov/ttn/amtic/files/ambient/pams/newtad.pdf.

U.S. Environmental Protection Agency (1999). A final draft copy of the "Particulate Matter (PM2.5) Speciation Guidance" Document, OAQPS, Research Triangle Park, NC 27711, October 7, 1999. http://www.epa.gov/ttn/amtic/files/ambient/pm25/spec/specfinl.pdf.

U.S. Environmental Protection Agency (2004a). National Air Quality Monitoring Strategy, OAQPS, Research Triangle Park, NC 27711, Draft April 2004. http://www.epa.gov/ttn/amtic/files/ambient/monitorstrat/allstrat.pdf.

U.S. Environmental Protection Agency (2003). Air Quality Criteria for Particulate Matter (Fourth External Review Draft), Office of Research and Development, National Center For Environmental Assessment, Research Triangle Park Office, Research Triangle Park, NC, EPA/600/P-99/002aD and bD., 2003.Volume I, Chapter 2, June 2003, http://cfpub2.epa.gov/ncea/cfm/recordisplay.cfm?deid=58003.

U.S. Environmental Protection Agency (2004b). List of Designated Reference and Equivalent Methods, August 26, 2004, NERL, Air Measurements Research Division, Research Triangle Park, NC. http://www.epa.gov/ttn/amtic/files/ambient/criteria/ref804.pdf.

U.S. Environmental Protection Agency (2004c). 2003 Urban Air Toxics Monitoring Program(UATMP), July 2004, Office of Research and Development, National Center for Environmental Assessment, Research Triangle Park Office, Research Triangle Park, NC. EPA-454/ R-04-003.http://www.epa.gov/ttn/amtic/files/ambient/airtox/ 2003doc.pdf.

Chapter 13

Autonomous Systems and the Sensor Web

Karl M. Reichard

Abstract

The sensor web is a collection of interconnected sensors and sensing systems sharing both raw and processed data with a collective capability much greater than that which could be achieved with a single dedicated sensor installation. The analogy is to the World Wide Web as a computer network versus any single computer network installed and operated by a single entity. If the goal of the sensor web for atmospheric science is the complete coverage of the globe, then fixed sensors and mobile sensor systems operated by humans will not be sufficient. Furthermore, we must consider the sensor web coverage area to be a three-dimensional space extending from the depths of oceans and other significant bodies of water to space in order to capture the drivers for atmospheric research. This paper describes issues related to the application of autonomous systems in building the sensor web. Key issues to be considered include levels of autonomy, defining requirements for autonomous systems, challenges in the design and deployment of autonomous systems, tactics for autonomously processing data collected from elements of the sensor web, and how to manage networks of autonomous systems.

Autonomous System Requirements

In considering the use of autonomous systems as part of the sensor web, it is instructive to begin by considering two questions:

- What attributes do we desire in a sensor system?
- What is an autonomous system?

Suppose we were to take out a classified advertisement looking for the ideal component to the sensor web: "Atmospheric researcher looking for sensor web component (AR looking for SWC)." What characteristics or attributes would we request? The "Wanted Ad" might include the following:

- Operates unattended
- Makes measurements continuously if necessary and lasts for a very long time
- Adapts to changing measurement requirements
- Is smart and can autonomously process sensor data and decide what future measurements are needed and what information to extract from the data
- Is low maintenance – should be able to operate for long periods of time, ideally forever, without maintenance
- Is a cheap date – must be less expensive than options using humans to collect, analyze, and manage data
- Cleans up after itself and manages data collected
- Likes new challenges (sensors)

We ideally are looking for a sensor system that collects and processes data with little or no input from a human operator and can change measurements, processing algorithms, update rates and other operating parameters in response to changing data analysis results or operating conditions. In addition, we want the sensor system to have a low total operating cost and be capable of adding new sensors as they become available or as the measurement requirements evolve over time.

Autonomous systems are generally assumed to be computer-controlled systems that possess human-like processing and reasoning capabilities. Characteristics of autonomous systems include

- Reaction to changing environment without human intervention
- Some level of decision-making capability
- Can receive instructions at high level (doesn't need lots of details)

Given these characteristics, there are two classes of autonomous systems we should consider for the sensor web:

- Autonomous sensors
- Autonomous vehicles

For both autonomous sensors and autonomous vehicles, the capability of the autonomous system can range from remote operation (limited autonomy), to capability mimicking human reasoning. While there is no general agreement on what capability the term "autonomous" implies, there have been attempts to define different levels of autonomous capability. One set of definitions is articulated in Table 1, where increasing levels of capability are assigned autonomous levels from 0 through 6, respectively.

Level 0 and level 1 autonomy as defined in Table 1 are fairly common. The higher levels of autonomy described in Table 1 are less common, but the listed attributes are beginning to appear in mission requirements within NASA, commercial, and military applications. The level of autonomy within a system should be chosen to match the system requirements since, in general, higher levels of autonomy require more processing resources and more onboard sensors (for operation), therefore increasing system cost and complexity.

TABLE 1: Levels of Autonomy.

Level	Attributes
6	Real-time multi-platform coordination
5	Adaptive response to faults and/or events – includes on-board replanning and response generation
4	Robust response to anticipated faults and/or events
3	Limited response to real-time faults and/or events
2	Execute planned mission with alternate plans
1	Execute preplanned mission without human in the loop (automatic control)
0	Remotely operated

What level of autonomy is needed in the sensor web? The sensor web is envisioned as a set of distributed nodes interconnected by a communications network that collectively behaves as a single, dynamic, adaptive observing platform. Autonomous systems provide adaptive capability – both in their processing capability and possibly mobility – and can automate the processing and interpretation of data from individual nodes, from subsets of nodes, or from the entire web. Clearly, the distributed nature of the sensor web means that it cannot be implemented with sensors that are all attended to by human operators; therefore, the sensor web requires at least autonomy level 1 as defined in Table 1. As the level of autonomy within the sensor nodes increases, the sensor nodes become capable of adapting to changing measurement demands and to changes in the nature of the information we wish to extract from the web.

In addition to geographically fixed sensors, the sensor web can also benefit from the use of sensors mounted on autonomous vehicles. There are several benefits to employing autonomous vehicles as sensor platforms:

- Geographic Access
  - — It is relatively easy to make measurements where there are people, but
  - — We need to make measurements where people are and where people do not live.
- Measurement Times
  - — We want to make regular measurements over long periods of time but also react to changing conditions.
  - — Using people to make measurements and manage measurement sites is expensive.
- Data Processing and Management
  - — We can easily collect more data than we have the ability to process manually.
  - — There is a need to fuse and access data from similar and dissimilar measurements.

For the same reasons that we cannot count on implementing the sensor web with fixed sensors constantly attended by human operators, we cannot expect to implement the sensor web with sensors mounted on tele-operated vehicles. In order to achieve global coverage with the sensor web, these mobile sensor platforms will require some level of autonomy beyond simple tele-operation. Greater autonomy will enable the platforms, to respond to changing measurement requirements and continue to provide valuable data.

Challenges in Autonomous Systems

There are several challenges to the widespread use of autonomous systems – not the least of which is the acceptance of systems capable of making their own decisions without human input. While there are certainly challenges in the areas of artificial intelligence that still need to be tackled to endow the autonomous systems with true intelligence, there are also many practical engineering problems that must also be resolved. If you ask almost any researcher in the fields of autonomous systems or remote sensor networks what their top three challenges are, they will probably respond "Power, power, and power." Electrical power is almost always a limiting factor in remote measurement systems. Most power for unmanned systems is provided by batteries. While they are getting lighter, more efficient, and more reliable, they still have a finite amount charge (operating time) and finite number of charge cycles (lifetime of measurement system). Eventually they will run out of power or will no longer be capable of being recharged and will need to be replaced.

Researchers are also working to develop environmental power sources capable of scavenging and storing power from the local environment. These include the use of geothermal sources, wind, waves, solar energy, and vibration. Nuclear power sources face cultural and safety barriers. Micro power plants are available that generate their own electrical power using much higher density sources (i.e., diesel fuel). While there are many alternatives to batteries, batteries are still the dominant source of power for remote sensors and unmanned systems. Renewable energy sources, while available, offer limited power and the available power will limit the types of sensors that can be deployed and the amount of local processing that can be implemented.

Another challenge is the level of intelligence that we can build into a sensor node or unmanned system. Even the "smartest" computer still does not have the ingenuity or problem-solving capability of a human. A colleague who worked for many years on the development of intelligent autonomous systems once remarked, "I would be happy if I could make an autonomous controller as smart as a bug, let alone trying to achieve human-like intelligence."

Automating sensor processing is easy compared to trying to make systems intelligent given appropriate computing resources and power. Nevertheless, the automation of sensor processing may still require significant *a priori* knowledge of expected measurement results. It may also require larges amounts of data to "train" algorithms. For example, researchers in pattern recognition and classification often refer to the 80/20 rule for classifier training and testing. In an ideal world, 20% of the available data would be used for training the algorithms and 80% used for testing the classification algorithms. Unfortunately, the ratio is often reversed due to the lack of training data.

While automating data processing is useful, what is really needed is autonomous processing. Building autonomous data processing systems capable of reacting to data circumstances that are new and deciding what processing is appropriate or necessary is still difficult. Anomaly detection can be used to detect data or situations that are outside the norm and trigger additional or conditional data processing, but all of these contingencies are typically programmed ahead of time.

Another challenge is data management. While it is often easy (and affordable) to acquire data, the problem of managing data can quickly become overwhelming. For example, consider the problem of temperature measurement. A single-point measurement of temperature requires a single value, but to accurately capture the temperature distribution over a one-dimensional space (e.g., temperature profile), two-dimensional, or three-dimensional space (e.g., sector scan) can require much larger amounts of data.

Processing data at the sensor can reduce the amount of data that needs to be saved. The problem becomes deciding if you really want to throw out the raw data measurements and only save the processed data. One possible solution is to save some of the raw data on a predefined schedule, but don't save all the raw data. Make the system smart enough to recognize data that is different and save the data that is new or different. The challenge is to make the system sensitive enough to recognize very slow changes in the environment and decide to save the data.

On board processing, embedded processing at the sensors, can be used to process data and convert the raw data to usable information. Processors, memory, and associated hardware are getting smaller, lighter, and less expensive. This solution still requires power to operate the embedded computing resources and can limit sensor or system life. Market forces are driving more innovation and improvement in communications than processors; hence, it's getting "cheaper" to transmit data than to process the data. This drives the need for more autonomy in the control of unmanned platforms and sensor systems so that communication resources can be used to transmit sensor data instead of for remote control.

An example where the local processing of sensor signals has been used to convert sensor "data" into useful system "information" is in the field of machinery health management [2]. The architecture shown in Figure 1 uses a distributed hierarchical architecture to process sensor data and convert it to information about the system health and condition. Local processing converts sensor data to features related to the condition of the specific subsystems being monitored. Processing is arranged in hierarchical layers to relate features from one subsystem to those from another and provide an overall assessment of the total system health and condition. The same type of hierarchical processing can be used with sensor nodes in the atmosphere where local processing is performed to extract relevant features (like a temperature profile measurement from a ground-based profilometery system) from several different "subsystems."

Another challenge is communication between sensor nodes and across the sensor network. The objective is to get the sensor data or processed information to the information users or customers. Communication channels are needed for data,

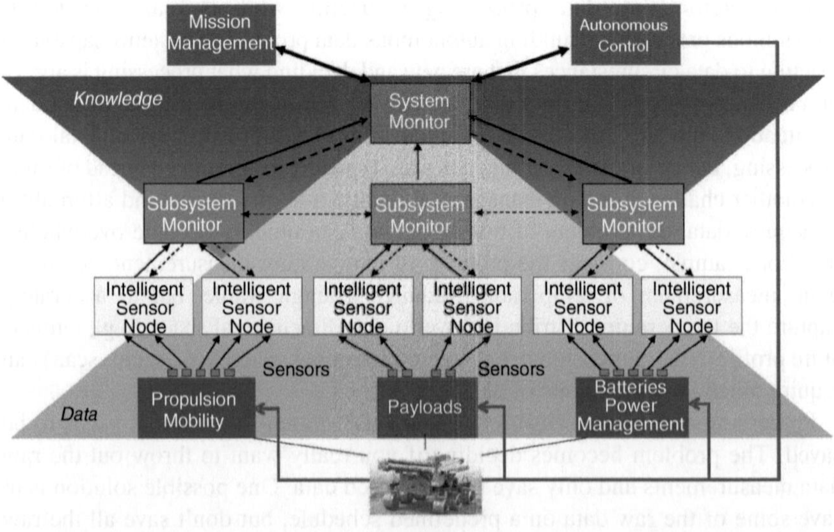

FIGURE 1. Processing architecture for assessing machinery health and condition.

programming, and control. Remote control or tele-operation can consume almost all available bandwidth and limit the ability to send back data. Sensor nodes can act as communication network nodes as well. This approach can improve network reliability, ensure connectivity, reduce power requirements, and leave wide bandwidth links (i.e., satellites) for control, if necessary. The use of sensor and network nodes that can be reconfigured on the fly to reroute data is known as mesh networking [3].

After the data are processed and saved on a server or other data repository, the challenge is providing access to potential users. What is needed are open standards and formats for communication and data that will allow different users or systems to access data from all potential sources. The question is how to request, read, and interpret data and information from a particular sensor web node? The trend is toward the use of XML for formatting data files. One advantage of XML is that it is text-based and is easily readable by humans. One disadvantage of XML, however, is that it can be inefficient for large data sets (i.e., imagery) because the format is text-based (that means that it requires a character for each digit in a number instead of representing the numbers in a more compact, binary format. Security can also be a concern in open systems but should not limit applications. For example, businesses use open standards for data and communication but require authentication for access to the data; thus, ensuring security. These same concepts can be used to ensure that only authorized users gain access to data.

One example of an effort to develop open standards for sensor networks is the open systems architecture for condition-based maintenance (OSA-CBM). This is a standardized architecture for system monitoring and condition-based maintenance systems [4, 5]. The OSA-CBM architecture breaks the monitoring system into

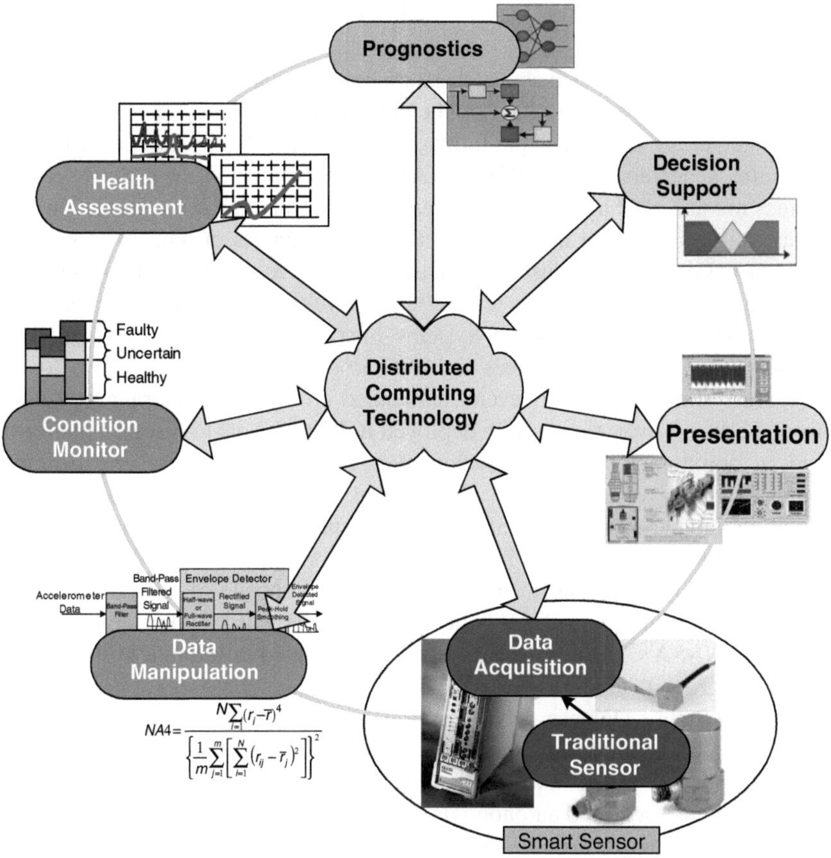

FIGURE 1. Open systems architecture for condition-based maintenance.

functional layers. The OSA-CBM standard defines the inputs and outputs required for each layer. Each layer may request data from any functional layer; however, data flow usually occurs between adjacent functional layers. With the inputs and outputs of each layer defined, the user, developer, or system integrator is free to use a variety of middleware approaches to implement the communication.

A final challenge in the implementation of sensor networks is time synchronization of the data. Time synchronization is particularly required for "array" measurements based on data from separate sensors or platforms. The required precision of time synchronization depends on the type of calculation being performed. Time synchronization is generally not a problem with more expensive and functionally capable systems (i.e., satellites), but it can be a problem in inexpensive sensor nodes or smaller unmanned vehicles. Larger systems such as satellites often incorporate accurate system clocks for applications such as GPS, but inexpensive processing systems and smaller unmanned vehicles often do not incorporate accurate clocks, so precise data synchronization can be more challenging.

Unmanned Platforms

Despite the challenges listed above, there are many unmanned systems in use today. Unmanned systems are classified by the domain in which they operate – ground, underwater, or air. Unmanned platforms are often referred to by the acronym UXV, where the X denotes ground (G), air (A), water surface (S), or underwater (U). There is not universal agreement or adherence to this naming convention, however, so care must be taken in interpreting references to classes of unmanned systems. Many professional and industrial societies have active organizations involved in promoting unmanned platform research and development. One of the largest is Autonomous Unmanned Vehicle Systems International (www.auvsi.org), which sponsors several large conferences each year around the world.

Figure 2 shows photos of several unmanned ground vehicles, ranging from a prototype Mars rover, to currently deployed military robots to research platforms used in autonomous system development. Figure 3 shows several different unmanned underwater platforms. Unmanned underwater vehicles range from miniature submarines capable of autonomous navigation, to unmanned surface vehicles (unmanned boats) to instrumented buoys that drift with the ocean currents carrying various sensor. Figure 4 shows a composite of several unmanned air vehicles: a small micro-UAV helicopter, propeller and jet-powered unmanned airplanes, NASA's EO-1 satellite, and an unmanned rotorcraft hosted by an unmanned ground vehicle.

The United State's Defense Advanced Projects Research Agency (DARPA), has sponsored a competition for unmanned ground vehicles, known as the DARPA Grand Challenge (http://www.darpa.gov/grandchallenge/). The 2004 DARPA Grand Challenge required autonomous ground vehicles to navigate a course from Barstow, California to Primm, Nevada, and offered a $1 million prize. None of the entries made it more than 7 km from the starting line and many did not

FIGURE 2. Unmanned ground vehicles (UGV).

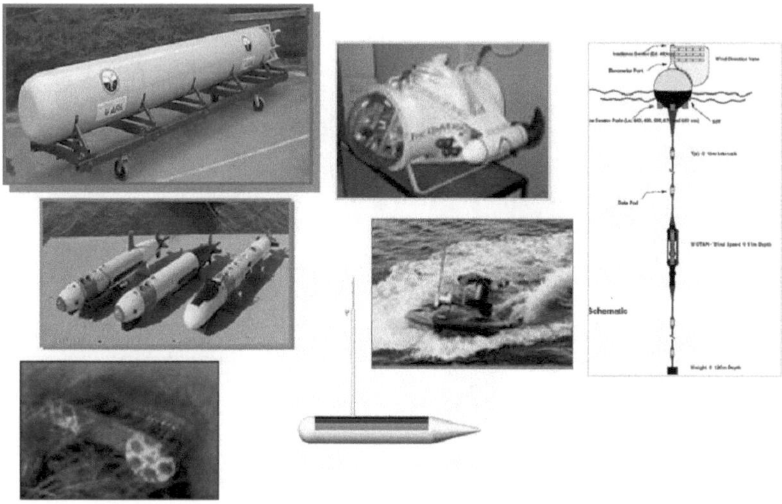

FIGURE 3. Unmanned underwater vehicles (UUV).

successfully complete the qualifying course. Figure 5 shows pictures of several of the entries from the 2004 Grand Challenge. A team from Carnegie-Mellon University in the US made it farther than any other team – 7 km. One of the keys to the success of the CMU team was the extensive use of satellite and airborne imagery to create detailed terrain maps and path plans for the area where the challenge course was to be located. The 2005 Grand Challenge will be held October 8, somewhere in the US desert southwest.

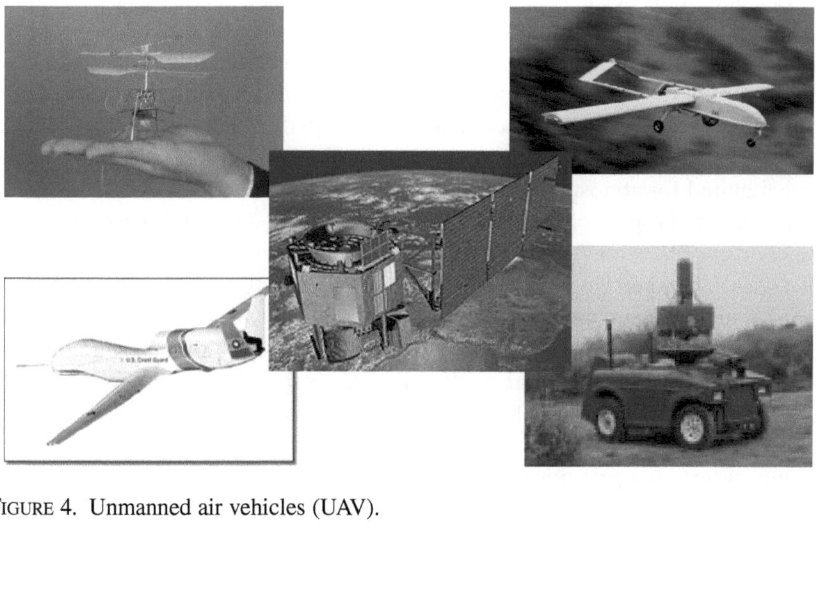

FIGURE 4. Unmanned air vehicles (UAV).

FIGURE 5. Several competitors from the 2004 DARPA Grand Challenge.

Networks of Autonomous Systems

The key to implementing the sensor web will be integrating data and information from networks of autonomous systems to make measurements from above atmosphere (satellites), in the atmosphere (UAV, balloons), beneath the atmosphere (UGV), and in the ocean (UUV, USV). It will be a challenge to make all of these systems work together. Collaborative control of autonomous systems is one of the leading challenges today for the autonomous control community. There is a need to ensure robust operation as nodes drop in and out of the network since we cannot guarantee continuous connectivity. Some networks of sensor nodes already exist as do sensor webs, but they have not yet been implemented on the global scale required for atmospheric science.

NASA's Jet Propulsion Laboratory and others have deployed sensor networks in a number of locations. Commercial industry has installed mobile and fixed sensor nodes for monitoring the condition of trucks, buses, trains, and airplanes to improve maintenance and reduce operating costs. Networks of acoustic sensors have been used to track changes in sound propagation due to variations in atmospheric conditions. Figure 6 shows some examples of sensors and sensor network applications that can be considered precursors to or potential elements of the sensor web.

Manufacturers are beginning to develop and market hardware specifically for sensor web applications. For example, the Intel Corporation and Crossbow

FIGURE 6. Examples of sensors and sensor network applications.

Technology, Inc. have combined to build and sell small smart sensors based on research conducted at the University of California Berkeley, called Motes [6]. The availability of off-the-shelf components for building sensor networks will certainly speed the population of larger sensor networks if communication issues between subnetworks can be resolved.

The Far Edge of the Sensor Web – Swarms

Many of the problems in controlling or coordinating multiple autonomous systems (whether they are vehicles or networks of sensors) result from needing to have larger amounts of information about the inner working of each autonomous entity. Autonomy researchers are currently looking to nature to study how animal species behave and coordinate their actions. One such area of research is swarming behavior.

The Marrian Webster Collegiate Dictionary provides the following definitions of Swarms:

1a (1): a great number of honeybees emigrating together from a hive in company with a queen to start a new colony elsewhere (2) : a colony of honeybees settled in a hive

1b: an aggregation of free-floating or free-swimming unicellular organisms—usually used of zoospores
2a: a large number of animate or inanimate things massed together and usually in motion : <swarms of sightseers> <a swarm of locusts> <a swarm of meteors>
2b: a number of similar geological features or phenomena close together in space or time <a swarm of dikes> <an earthquake swarm>
3: to form and depart from a hive in a swarm
4a: to move or assemble in a crowd; 4b : to hover about in the manner of a bee in a swarm

Bruce Clough from the US Air Force Research Laboratory has described Swarms as, "A bunch of small cheap things doing the same job as an expensive smart thing." This definition, while not as encompassing as what we get from the dictionary, captures the motivation for the study of swarms by the autonomy community – reduce the cost and complexity of collaborative autonomous systems by creating them from a collection of simple, cheap, components governed by a few simple rules.

For example, consider a school of fish. The simple rules for the fish might include

1. Head toward the center of mass (of the school)
2. Swim in the same general direction
3. Don't run into your neighbors

Swarms react to external events without prepared plans and without explicit models of how the collection of individuals interacts with the environment. We see examples of swarming behavior in plants and lower animals. Manmade systems that try to mimic swarm behavior typically use local control of the swarm member with feedback from the actions of other swarm members, but tend to use less memory and computing resources than collaborative systems with full models of what every member is doing. Some of the characteristics of swarms include

- Simplicity
  — Complexity Is Emergent
- Resiliency
  — Attrition Tolerant
  — Individual imperfection okay
- Scalability
  — Adding Members Easy
- Probabilistic
  — Emergent Behaviors Are Stochastic, Not Deterministic

Where do swarms fit into the sensor web? There is interest in the use of swarm principles in the design of control systems for various unmanned platforms because the control implementations tend to be low cost, simple, and easy to replicate. The problem is that you can't really tell what a swarm is going to do before you watch the behavior emerge, which can be costly if the behaviors that emerge cause damage to the unmanned platform or other systems. Swarms may provide an interesting concept for controlling which measurements a group of sensor nodes make – particularly when they are just "sitting around" waiting for something to happen. Emergent computing behavior could provide a solution to short-term needs for more computing capability than a single sensor node might have. For example, genome processing using distributed network of desktop PCs can be considered a derivative of swarm behavior. The real application may be in the use of a hybrid of swarm and team behavior. Consider the example of a football team (American or soccer) where the game is governed by rules and plays, but ultimately depends on the instinctive response and swarm behavior of the players to score.

Conclusions

In order to implement the sensor web, we will need to use systems that don't require constant human operation or oversight. When considering the application of Autonomous Systems, we should remember that they include not just platforms that move, but also fixed sensor systems that can adapt and react to their own changing environment. One caution in autonomous systems is that we must be careful what we wish for because the system might start doing things we don't want or didn't anticipate. One of the greatest expenses in the development and deployment of autonomous systems is system verification and validation. Finally, the sensor web is already here – many of the pieces exist. The challenge in creating the sensor web will be in learning how to connect the nodes and access the information.

Acknowledgements This paper is based upon work supported by NASA under the Engineering for Complex Systems Program – Autonomous Propulsion Systems Technology Subprogram (Grant No. NAG3-2778).

References

1. B. Clough, Metrics, Schmetrics! How Do You Track A UAV's Autonomy?, Proceedings of the 1st AIAA UAV Conference, Portsmouth, Virginia, May 20-23, 2002. Also, Unmanned Aerial Vehicles Roadmap 2000-2025. Office of the Secretary of Defense, Washington DC. April 2001.
2. K.M. Reichard, E.C. Crow, Self-Awareness, Monitoring and Diagnosis for Autonomous Vehicle Operations, Proceedings of AUVSI Unmanned Systems 2002, Orlando, FL, 2002.
3. Mesh Radio Network Performance in Cargo Containers, *Sensors Magazine*, March 2005.
4. www.osacbm.org.
5. M. Lebold, K. Reichard, P. Hejda, J. Bezdicek, M. Thurston, A Framework for Next Generation Machinery Monitoring and Diagnostics, Proceedings of the 56th meeting of the Society for Machinery Failure Prevention Technology, Virginia Beach, VA, April 2002.
6. Smart Sensors to Network the World, *Scientific American*, June 2004.

Chapter 14

Comparison of Measurements—Calibration and Validation

PAUL A. NEWMAN

Cal/Val and the Cal/Val process

Why Is Cal/Val Important?

Science is founded upon its theories and hypotheses. However, those theories ultimately require a solid numerical foundation. We ultimately find that the better the numbers, the better the science. Of course, these numbers ultimately derive from observations, and observations depend on instruments. The importance of calibration and validation (cal/val) comes from the need to (1) quantify scientific statements, (2) combine instrument observations, (3) analyze long-term behavior of systems (trends), and (4) provide inputs for other science applications.

This science ultimately begins with a question. Simple science questions can answered with simple instruments. For example, determining the annual cycle of temperatures at a specific point requires a simple mercury thermometer with a lettered scale. Since the annual cycle is about 30 °C in the Washington, DC, region, a 1 °C thermometer error is only a 3.3% error of the annual cycle. However, if a 10-year trend is needed to better than 1% accuracy, than a more stable thermometer is needed. The instrument's accuracy must be better than 0.3 °C over a 10-year period. Hence, this thermometer needs to be calibrated on at least an annual basis. If the trend is required over a large region, for example, the East Coast of the United States, then a number of thermometers are needed. This group of thermometers would require consistent cross-calibration, consistent mounting arrangements, and a movable standard. Complex science questions require more complex calibration and proof that the calibration is not shifting. The answer to our science question must have a foundation of proven observations.

Temperature observations are also required for numerical weather prediction (NWP) and for analyzing weather phenomena. First, the vertical temperature profiles provided by regular meteorological balloon observations (radiosondes) are critical for accurate forecasts of synoptic scale weather systems (Graham et al., 2000). Furthermore, the accuracy of these temperature observations is a limiting factor in the accuracy of NWP. Errors in forecast initializations are propagated by the forecast model, either masking actual synoptic systems or developing false systems. Consistent and calibrated temperature observations reduce NWP problems. In this case, the quality of the observations has a direct social impact.

Our simple temperature example above begins to expose the complexity associated with the simple question, “Is Earth’s atmosphere warming?” This is not the question of whether climate change is connected to CO_2 increases, but the simple question of whether our observations show that our atmosphere is warming. Temperature observations of the atmosphere are now principally derived from 2 sources: radiosondes and satellites. Prior to the 1970s there were no satellite observations, and prior to the 1940s there were few balloon observations. The radiosondes present a particular problem because they were originally designed to provide simple observations for weather forecasting, rather than climate trends. For a globally averaged temperature, we need to combine radiosonde observations, yet there are substantial differences in radiosondes in both the measurement and the operations [e.g., see Ivanov, 1991]. Radiosonde manufacturer changes led to significant temperature shifts that compromise trend estimates from single stations [Gaffen et al., 2000]. While our temperature observational systems may be great for analyzing a large number of science questions, they present a serious challenge for long-term trends because of the absence of good calibration and validation.

Scientists are occasionally guilty of rushing to make an observation without considering cal/val properly. By experience, scientists know that bad data are usually worse than no data at all. In 1958, poorly calibrated spectrographic plates were used to make observations of the total amount of ozone over Antarctica. These observations were eventually published as evidence that an ozone hole had been over Antarctica in 1958 prior to the introduction of large quantities of chlorine based ozone depleting substances [Rigaud and Leroy, 1990]. Subsequent analysis showed that these observations were inconsistent with the better calibrated 1958 Dobson spectrometer observations from Antarctica, inconsistent with the meteorology in 1958, and inconsistent with the ozone hole patterns observed in the 1990s [Newman, 1994]. The poor spectrographic plates led to a waste of time, money, and energy.

The Cal/Val Process

Every instrument has a theoretical background that leads from a voltage/current/resistance measurement (hereafter, referred to as voltage for simplicity) to a geophysical observation. Ultimately, instruments only record voltages. For example, thermistor observations of temperature can be made because the resistance of the thermistor is proportional to the temperature. The thermistor measurement is a practical application of material science theory. This voltage is ultimately turned into a geophysical quantity by a calibration process that is either predetermined or is applied in real time by measuring targets with known temperatures. The thermistor illustrates that every measurement has a theoretical basis.

The process of calibration/validation is illustrated in Figure 1. The first step in making an observation begins with the measurement of a voltage. Step 2 involves

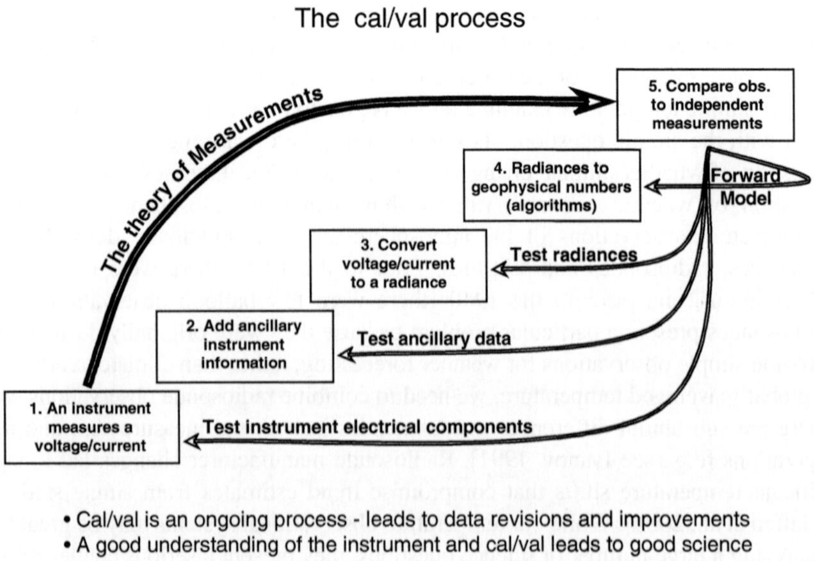

FIGURE 1. The Cal/Val process.

adding ancillary information to our measurement. This ancillary information may involve calibration information that is determined from the lab rather than in the field. It may also involve observations of conditions that render the observation invalid, such as stray light for optical instruments, and contamination for in-situ instruments. The third step in the conversion of the voltage/current/resistance into a radiance. The fourth step involves the algorithm that converts a radiance into a geophysical quantity such as total ozone. In our example of the thermistor, this fourth step does not exist. The final step of the cal/val process is the comparison of the geophysical quantity to an independent observation. Of course, at each step, it is necessary to test the results. For example, in step 1 it is always necessary to monitor the electronics of the instrument. In step 2, ancillary information from the lab on an instrument's nonlinear performance is crucial for understanding its behavior. Step 3 involves a comparison of radiances from an independent instrument. An example would be the comparison of spectra from two multi-spectral instruments. Step 4 requires thorough tests of the algorithm. An example of this would be the comparison of the UV-vis spectrum of NO_2 to a spectrum computed from an actual NO_2 profile (an example of forward modeling). The entire process requires continuous monitoring and cross checks for a careful observation.

This paper has separate sections on calibration and validation. Calibration is discussed first. In this section calibration is defined, some general principles of calibration are derived, and some problems associated with calibration are illustrated. This calibration section also includes a very brief discussion of algorithms. Validation is discussed later.

Calibration

What Is Calibration?

Calibration is defined as the quantification of an instrument's response to a known input. However, how do find a known input? An example is the measurement of the height of a person. The meter stick is derived from known standards that were originally kept as platinum bars. These bars were supposed to be 1/10,000,000 of the distance from the Equator to the pole. The meter is now defined as 1,650,763.73 vacuum wavelengths of light resulting from unperturbed atomic energy level transition $2p_{10}$ - $5d_5$ of the krypton isotope having an atomic weight of 86. All fundamental units of measurement are now established by the International System of Units. Calibration is an ongoing process. It begins with the first design of the instrument. The design must ensure that calibration capabilities are traceable against known standards and can be performed over the lifetime of the instrument. A complete error analysis is usually performed to test whether the signal to noise is adequate. Measurements are usually built into an instrument (e.g., temperature, pressure, etc.) that keep tabs on the instruments stability and health. In addition, calibration is done for both instrument subsystems as well as the instrument output.

A calibration curve is schematically shown in Figure 2. The solid black curve is defined by a series of source inputs ranging from 0–5 on the x-axis. The instrument returns a voltage of 0–9 volts. The intercept (B) is conventionally called the "dark current." The slope of the curve is the gain (G). Linear behavior is highly desired in an instrument because only 2 calibration points are required to determine the output of the instrument. A strong gain is also highly desired because this means the instrument is very sensitive to small variations of the input.

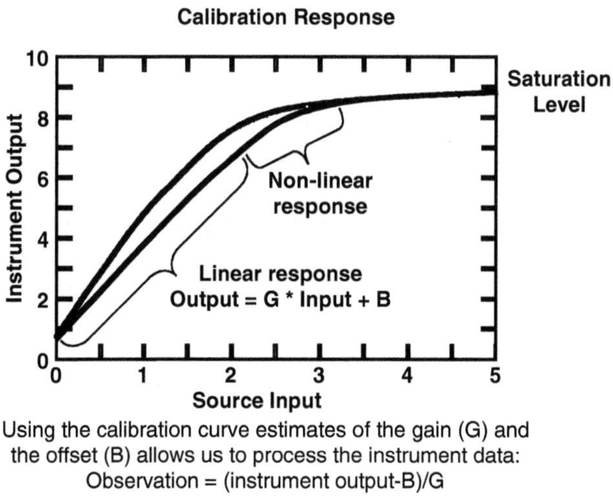

FIGURE 2. Calibration curve for a hypothetical instrument.

Calibration curve problems are also illustrated in Figure 1. Saturation occurs when an instrument stops responding to an increasing input. Non-linearity and saturation are always found in instruments and are both undesirable quantities. The nonlinear regime requires greater numbers of calibration points to calculate the instruments output. Another undesirable aspect of instrument response is hysteresis. The dashed line in Figure 2 shows a hysteresis effect that occurs when the instrument detector saturates. As the input decreases, the output remains artificially high and doesn't return to its original state until the source input goes to zero in this example. An additional problem found with calibration curves is a time trend. Because of detector degradation and other problems, the gain, dark current, saturation, and nonlinearity may change as an instrument ages. To prevent drift in observations, the calibration curve requires regular updates. Instrument degradation is an extremely common problem that seriously compromises attempts to calculate trends in time.

Calibration Examples

Example 1: Carbon Dioxide

Carbon dioxide has been increasing in our atmosphere since the start of the industrial revolution in the 1800s. The quality of the CO_2 measurements provides the foundation for our current assessments of climate change. In-situ observations of CO_2 are made from a number of stations around the globe, and these stations can be used to decipher sources and sinks of CO_2 [Tans et al., 1990]. The CO_2 instrument uses absorption by the strong CO_2 line at 4.26 μm. Air is drawn into the instrument and passed between the lamp and detector. At this wavelength, the amount of IR absorption is directly proportional to the CO_2 concentration.

Figure 3 shows the calibration curve for a series of injections of air with known concentrations of CO_2 for the instrument at Argyle, Maine, on May 2, 2004. The CO_2 amounts were predetermined gravimetrically and were shipped to the station in gas bottles. The instrument samples continuously. However, at regular intervals during the day, the calibration gases are passed into the instrument. As is apparent from the line, more CO_2 results in less detection of IR light. The instrument exhibits some small non-linearity as is seen from the fit to the 4 points on the curve.

These CO_2 observations provide the basis for determining that greenhouse gases have been increasing in our atmosphere. Our current assessments of climate change that provide information for our public policies are founded on this high quality data that has a well proven foundation.

Example 2: Brightness Temperature from the Advanced Microwave Sounding Unit (AMSU)

The Advanced Microwave Sounding Unit (AMSU) is a 20-channel microwave radiometer that is placed aboard the NOAA polar orbiting satellites to provide temperature and humidity information [Diak et al., 1992]. AMSU is actually two

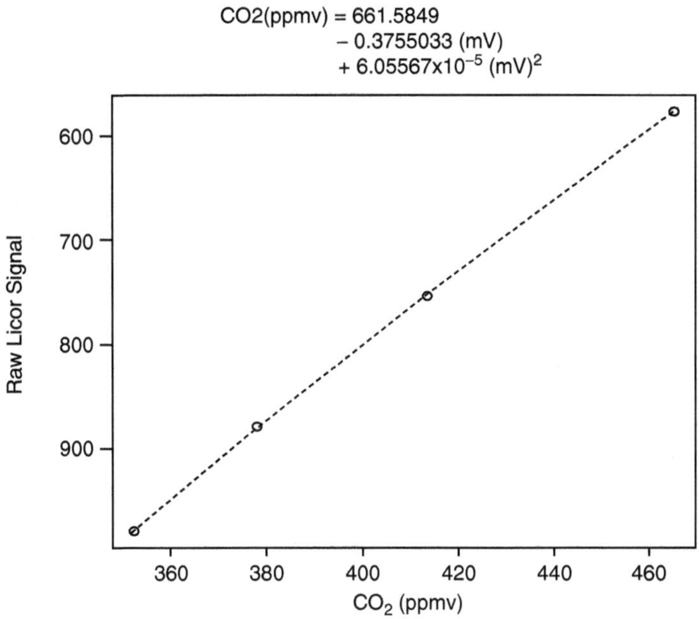

FIGURE 3. CO_2 calibration curve for observations at Argyle, Maine, on May 2, 2004. The 4 black points are from WMO CO_2 gas standards. Figure courtesy of Dr. Arlyn Andrews (NOAA/CMDL).

separate instruments: AMSU-a is a 15-channel radiometer for measuring temperature, and AMSU-b is a 5-channel radiometer for measuring humidity profiles. AMSU-a is further split into two separate instrument modules: AMSU-a1 with 13 channels and AMSU-a2 with 2 channels. The advantage of microwave radiances is that they are transmitted through most clouds. The AMSU-a radiances are used in inversion algorithms to provide temperature profiles for NWP. The AMSU-a radiance observations at 57.95 Ghz are directly proportional to the layer mean temperature of the upper troposphere and lower stratosphere.

The other AMSU channels provide temperature information from the surface to approximately 3 hPa. The scanning capability of AMSU combined with the polar orbit leads to near complete day/night global coverage of the Earth's atmosphere at a resolution of less than 100 km. Hence, AMSU provides a direct data set for evaluating climate change of nearly the entire atmosphere.

The calibration of AMSU-a is documented in the NOAA KLM User's Guide (http://www2.ncdc.noaa.gov/docs/klm/index.htm). During each 8-second AMSU-a scan, the instrument makes 30 Earth observations, and view 2 internal targets twice: a view of the cosmic background (2.72K), and a view of an internal target (~300 K). The preflight ground calibration is used as ancillary information to provide the nonlinear response of the instrument. The radiance of the Earth scene is determined from the internal target counts, the cold space counts, and a nonlinear correction that is determined from the preflight calibration. The cold space counts and internal target counts are actually estimated from a running average of several adjacent scans. The internal target radiance is calculated from the Planck function using an average of several platinum resistance thermometers (each of which has its own ground calibration coefficients). The cold space radiance is known and the target radiance is calculated from the target temperatures. This 2-point calibration line is then used to convert the Earth view counts into a black-body radiance.

Calibration Problems & Uncertainties

Every instrument has calibration problems and uncertainties. The AMSU radiance example provides an excellent example of calibration uncertainties. For example, the internal target presents a number of questions. The exact emmissivity of the internal target and the internal target temperature both need to be known with great accuracy and precision. The target is monitored with platinum thermistors, and these need to be known with great accuracy and precision. These thermistors are calibrated on the ground and may change over the life of the instrument in space. Hence, the instrument's main calibration point (the internal target) is critically dependent on the calibration of the thermistors, which, in turn are dependent on laboratory standards and thermistor stability. The cold space view much also take into account some background variation and contamination from the Earth's limb and spacecraft parts. Particular attention needs to be paid to the stability of the instrument's frequencies and to the nonlinearity of the instrument. Another problem with AMSU is the lack of a third point for calibration. After prelaunch calibration to assess nonlinearity, it is extremely difficult to determine drift in the nonlinear behavior of the calibration curves.

The instrument uncertainties are also critically dependent on the standards. However, all instrument uncertainties are only identified if the instrument can be compared to some other measurement of the same quantity. This requires validation.

Algorithms

Before jumping to validation, it is necessary to discuss algorithms. An algorithm is a procedure that takes the natural measurement of an instrument and converts it into a geophysical number. In our AMSU example, the instrument measures a black body radiance. These radiances are then converted into temperature profiles. If the algorithm is flawed, then even a perfect radiance can be ruined.

As with instruments, algorithms involve a series of physical assumptions that require lab work, calibration, testing, and validation.

A simple algorithm example comes from ultraviolet (UV) absorption by ozone (Figure 4). The absorption of the incident flux (I_0) is attenuated as it passes

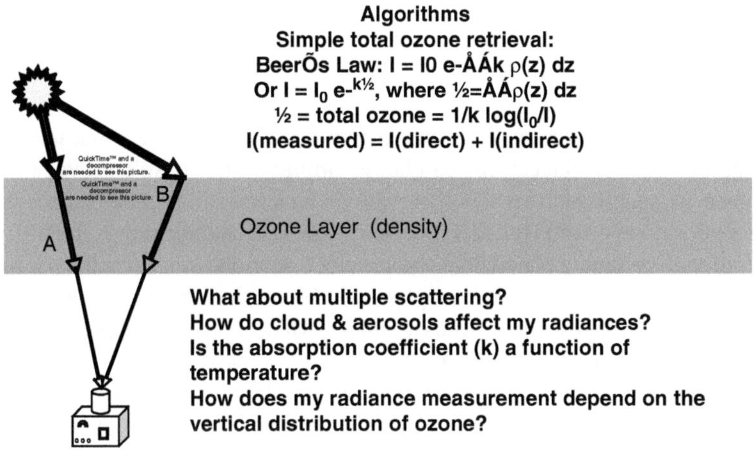

FIGURE 4. Schematic example of algorithm and possible problems with conversion of a well-calibrated radiance to a geophysical quantity.

through the ozone layer along path A. The expression for the absorption is shown in Figure 4, where k is the absorption constant, ρ is the ozone density, z is altitude, I is the measured flux, and Ω is the integrated column ozone amount. In principle, the total ozone is a simple relationship of the absorption constant (k), the measured flux at the ground, and the incident flux at the top of the atmosphere.

This algorithm assumes that the incident flux includes just the direct beam. If it doesn't, then we need to include an indirect scattering correction (along path B). We also assume that clouds and aerosols do not affect our flux. This is wrong, because aerosols do absorb and scatter in the UV. We also assume that k a constant as a function of altitude. This is also wrong, since k varies with temperature, and temperature is not a constant with altitude. Furthermore, we assume that the measurement does not depend on the vertical distribution of ozone, but this is also wrong, particularly for large UV flux attenuation. While our simple instrument might give excellent actinic flux measurements, the algorithm correction to total ozone is flawed.

Our total ozone algorithm example illustrates key points. First, a perfect observation of a flux does not guarantee a perfect observation of a geophysical quantity. Second, an algorithm always contains a number of assumptions. Each assumption must be tested and validated to test the full algorithm. In our example, some simple theoretical work and a forward model could be used to test our assumptions. Finally, even a perfect ozone algorithm must consider other factors that may influence a flux or radiance (e.g., aerosol absorption, NO_2 absorption, solar variations, etc.).

Validation

What Is validation?

Validation is the comparison of an instrument's measurement to some independent measurement(s). Successful validation implies (but doesn't prove) that an instrument is sound and that the observations have a solid and defensible basis, while a poor comparison (failed validation) exposes fundamental problems in an observational system. As noted in the previous section, even a perfectly sound instrument may produce flawed geophysical observations through a poor understanding of the physics of the measurement. Hence, validation is an absolute necessity for a scientific instrument.

Validation can be classed into two categories: internal and external calibration. Internal validation is the process of using identical instruments to measure the same quantity. Some instruments alternate detectors or channels to ensure that the instrument is internally consistent. A good example of internal calibration is the use of different wavelength pairs by the SBUV instrument to produce a total column ozone observation. External validation uses a completely separate instrument's data for comparison. An example of external validation is the comparison of the Total Ozone Mapping Spectrometer (TOMS) satellite instrument observations to ground-based Dobson spectrometers. External validation can be further partitioned into a direct comparison of observations that come from the instrument and its algorithm, and validation of radiances by the use of forward models. An example of radiance validation is the computation of the UV spectrum observed by SBUV as derived from an ozonesonde profile and the incident solar flux.

There are numerous requirements for validation. First, validation begins with the documentation of the instruments. This is a critical need, since the instrument heritage, the calibration techniques, and the algorithms must be clearly understood before comparisons are performed. Second, validation requires an independent evaluation of the instrument design and data analysis scheme. Third, blind (or double blind) instrument and data analysis intercomparisons must be performed (Figure 5). Fourth, forward models must be used consistently to validate radiances. Fifth, progress and actions must be determined and passed back to the experimenters. Sixth, validation must be an ongoing activity.

Validation is usually quite different for different types of instruments. As examples, validation plans for various instrument types have been developed by the Network for Detection of Stratospheric Change (NDSC). The NDSC is a set of high-quality remote-sounding research stations for observing and understanding the physical and chemical state of the stratosphere and how the stratosphere might be changing. Since the NDSC is distributed around the globe and since they need to establish stratospheric trends, the NDSC has an extensive and ongoing cal/val program to insure data quality for long-term trends (see www.ndsc.ws). A number of validation plans and efforts have been conducted by the NDSC for a variety of instrument types.

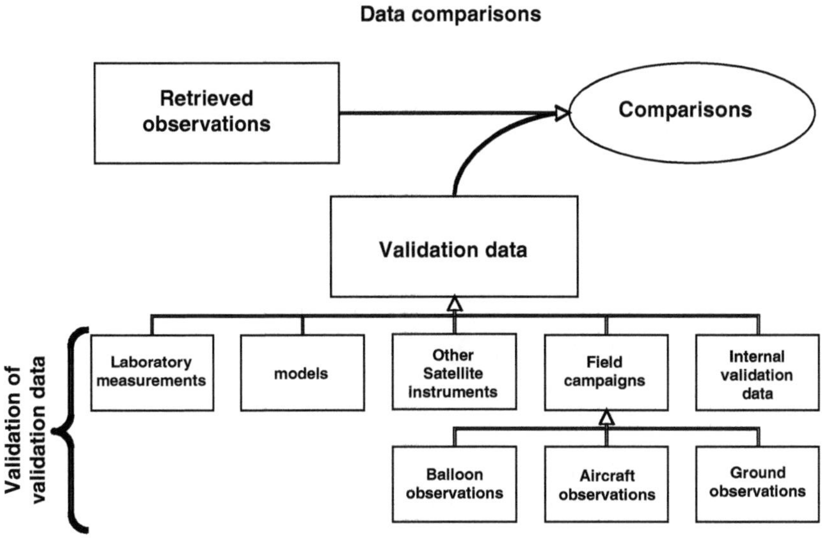

FIGURE 5. Sources of validation data.

Laboratory Validation

A validation effort begins with a broad plan that extends from the lab to atmospheric observations. Lab validation uses observations by multiple instruments in controlled circumstances. This typically occurs during an instrument development phase and continues with field comparisons. An excellent example of lab validation is the NDSC study of UV spectrometers that measured atmospheric NO_2 [Hoffman et al., 1995]. Seven spectrometers were compared in the lab by measuring the 435.8 nm Hg line from a calibration lamp (see Figure 6).

From such a lab intercomparison, the actual instrument spectral resolution, wavelength offsets, and wavelength asymmetries were determined. These differences are extremely important when the absorption cross-sections are used in algorithms to derive the NO_2 column amount.

Internal Validation

Internal validation is performed by measuring a quantity in an instrument with two separate processing streams or channels. Internal validation is useful for testing for instrument drift and stability. Such a technique usually begins with the initial design of an instrument. In most cases, weight, volume, power, or cost prevents the design of an instrument with multiple channels. Nevertheless, internal validation has proven to be an extremely useful technique for characterizing instrument behavior, especially for satellite instruments that can't be brought back to a lab for recalibration.

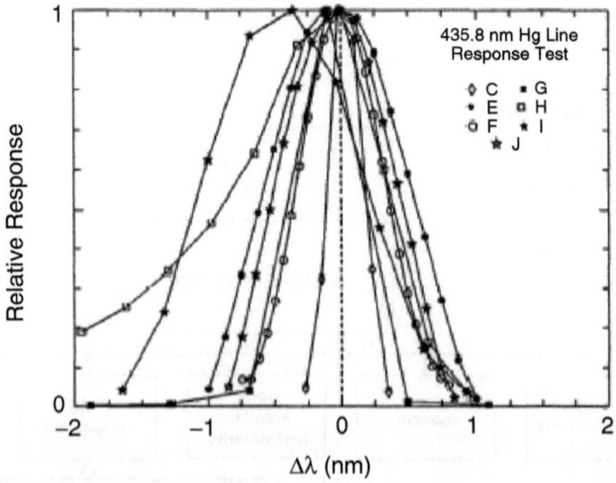

FIGURE 6. Response of 7 different UV spectrometers to the 435.8 nm Hg line from a calibration lamp. Adapted from Hoffman et al. [1995], Figure 5.

An example of internal validation comes from the NOAA Solar Backscatter Ultraviolet Spectrometer (SBUV-2) instrument on the NOAA-9 polar orbiting satellite. This instrument measures total column ozone globally every day. The measurement uses wavelength pairs to calculate column ozone. In Figure 7 column ozone is shown for the difference between the A-pair and B-pair column estimates before (red line) and after (green line) calibration correction by D/B Pair method for NOAA-9 SBUV/2. The new algorithm shows marked improvement with a reduction in the bias between the two estimates, less month-to-month variability, and less long-term drift between the two observations.

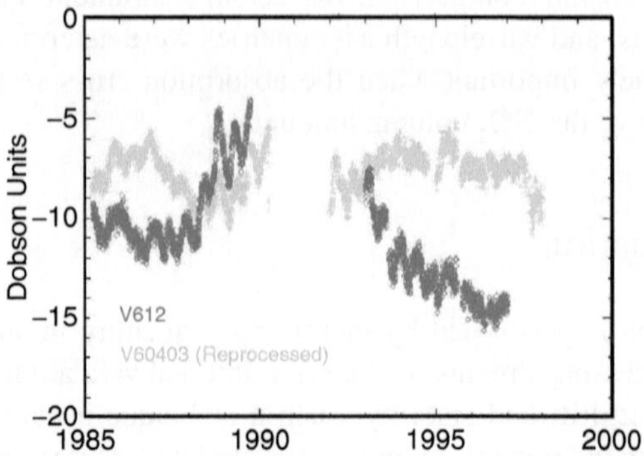

FIGURE 7. Plot of A/B Pair differences before and after calibration correction by D/B Pair method for NOAA-9 SBUV/2. The D/B Pair comparison provides an estimate of instrument change. The A/B Pair serves as a check on the improvement from these corrections. Provided by S. Kondragunta NOAA/NESDIS.

External Validation

External calibration is best done when instruments are viewing the same air mass and is best accomplished in organized coordinated campaigns (e.g., the NDSC NO_2 comparisons). However, oftentimes an organized campaign is not available. In the case of satellite instruments, many rely upon coincident observations from satellite overpasses of stations that make regular daily observations. Over the course of a year, a sufficient number of "coincident" measurements are obtained to statistically characterize the satellite-ground instrument differences.

An example of simultaneous in-situ observations come from the SAGE III Ozone Loss and Validation Experiment (SOLVE) that was conducted from Kiruna, Sweden, in the winter of 1999/2000 (Newman et al., 2003). In this campaign, the NASA ER-2 high altitude aircraft carried 4 different N_2O instruments. Figure 8 displays the instrument comparison as adapted from Figure 2 of Hurst et al. [2003]. The ACATS instrument is a gas chromatograph, ALIAS and ARGUS are tunable diode lasers, and WAS is a whole air sampler. This intercomparison revealed that N_2O measured between instruments usually differed by only about 3–4% [Hurst et al., 2003].

Remote sensing instruments are again best validated by viewing the same air mass. As was discussed earlier, the NDSC comparison of UV spectrometers that measured NO_2 was first conducted in the lab. The second step involved simultaneous observations of NO_2 slant columns [Hoffman et al., 1995]. A second example of validation comes from the comparison of lidar instruments during the second SOLVE mission that was flown in January 2003. Figure 9 displays the ozone difference between the 2 lidars between about 10 and 26 km. The lidar ozone observations reveal that the AROTAL lidar is higher than DIAL at lower altitude (12–18 km), but lower than DIAL from (18–26 km). The individual points are shown in the figure as yellow dots.

Satellite radiance validation

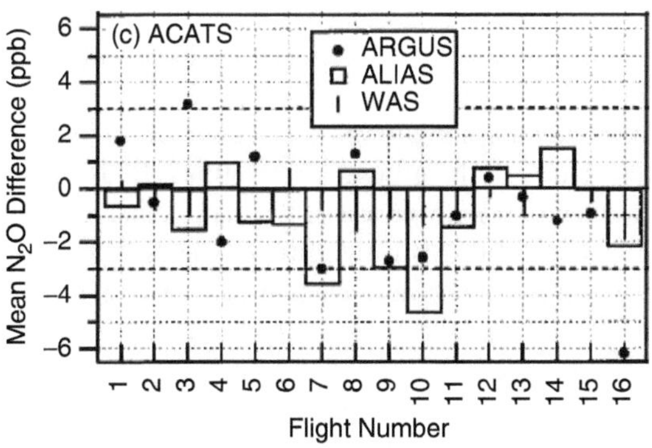

FIGURE 8. Percentage difference of in-situ N_2O observations from the ARGUS, ALIAS, and WAS instruments with respect to the ACATS instrument aboard the NASA ER-2 during the SOLVE experiment. Figure adapted from Figure 2 of Hurst et al. [2001].

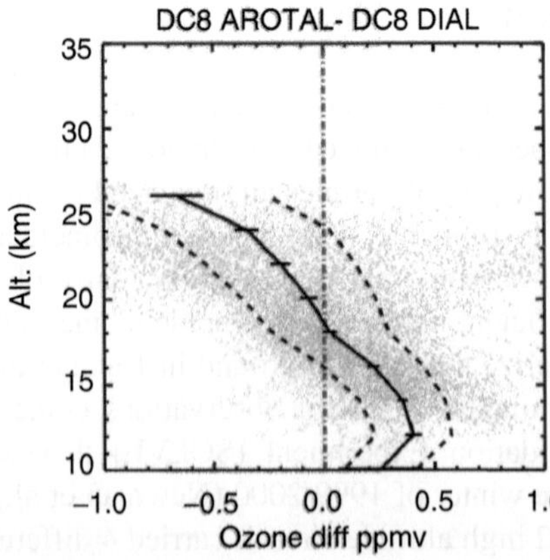

FIGURE 9. Comparisons between AROTAL and DIAL ozone lidars during SOLVE-2. The thick line is the mean difference at the different altitudes. The short horizontal lines indicate the 90% confidence limits of the mean. The dashed curves show the std dev of the population of differences. Figure from L. R. Lait et al., 2004.

Satellite instruments are rarely recovered after launch. Typical satellite orbits are higher than the space shuttle can fly, so servicing or retrieval of an instrument is extremely difficult. Furthermore, the science return from the recovery of an instrument on a low orbit satellite must be carefully measured against the safety of the astronauts and the costs of recovery. Once launched, satellite instruments can't be recalibrated, repaired, or modified. Hence, they require careful monitoring to determine their stability and possible drifts in the measured radiances.

An example of a satellite instrument's degradation comes from the Nimbus-7 TOMS instrument. During the 1980s, it was found that the global trend did not match the trend determined from Dobson ground instruments. Analysis of the instrument revealed that the diffuser plate had degraded as a result of extended exposure to the sun. This led to an underestimate of the incident solar flux that translated into erroneously low total column ozone observations. The TOMS algorithm was then modified to cancel the diffuser plate effect on the total ozone estimate. This then led to a substantially improved column ozone estimate that was in much better agreement with the ground observations. In this case, the instrument problem was eliminated by a clever use of the data in and improved algorithm.

A second method for improving satellite validation is to choose observation targets that have well-known characteristics. In the case of TOMS, the instrument can be used to measure the reflectivity of Hudson Bay. During winter, the Bay freezes and snow is generally blown off the surface. This leaves a uniform surface for the satellite to observe on cloud-free days. During summer, the thawed surface also provides a relatively uniform ice-free surface. Long-term drift of the instrument can now be monitored by year-to-year comparisons of the reflectivity. Figure 10 displays the reflectivity of Nimbus-7 TOMS at 380 nm from late-1978 to 1993. The small changes in this "reflectivity" show that the instrument drifts

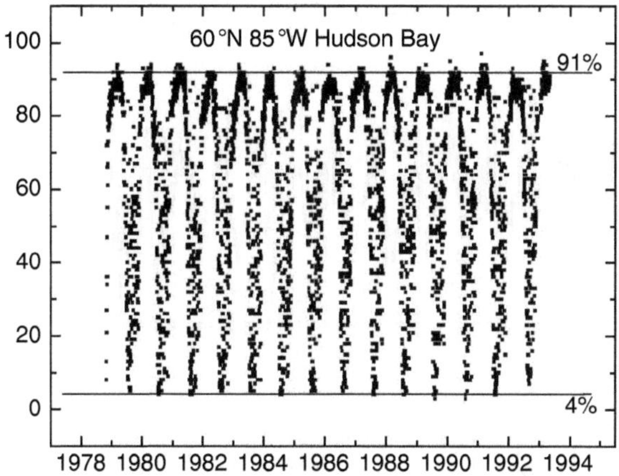

FIGURE 10. Internal TOMS instrument validation from viewing of a known target (Hudson Bay) at 380 nm. Year-to-year variations are small, showing the long term stability of the TOMS instrument over this period. Figure courtesy of J. Herman (NASA/GSFC).

have been adequately accounted for in the radiance calibrations. Viewing of known "external" targets can be used to either calibrate or validate an instrument. In this case, the 380-nm reflectivity provides a validation of the radiances.

Satellite Validation by Ground Observations

Stratospheric ozone depletion has been assessed on a regular basis over the last 30 years. The quality of the observations is key to those assessments. The comparison of TOMS total ozone to Dobson spectrometer observations has provided a solid foundation for the "believability" of the observations. Figure 11 shows a comparison of TOMS to the Hohenpeisenberg Dobson observations. The upper portion of the panel shows the Dobson data, while the lower panel shows the percentage differences between TOMS and the Dobson. The large interannual variability of the Dobson is captured by TOMS since the difference shows no annual cycle. Further, little drift is found between the two data sets over the extended period.

The TOMS observations also provide a useful transfer standard for the Dobson stations. TOMS provides daily global coverage, so it is possible to compare individual Dobson instruments using TOMS. For example, the Hohenpeissenberg can be compared to Brisbane, Australia, by subtracting TOMS from both data sets, revealing subtle changes in the Dobson instruments. Figure 12 shows the TOMS comparison to Brisbane. In February 1985, the Brisbane Dobson versus TOMS shows a rapid 8% downward shift. This shift is not apparent in the Figure 11 Hohenpeissenberg data or other stations, suggesting that the shift is unique to Brisbane. Station records indicated that the Brisbane Dobson was automated in

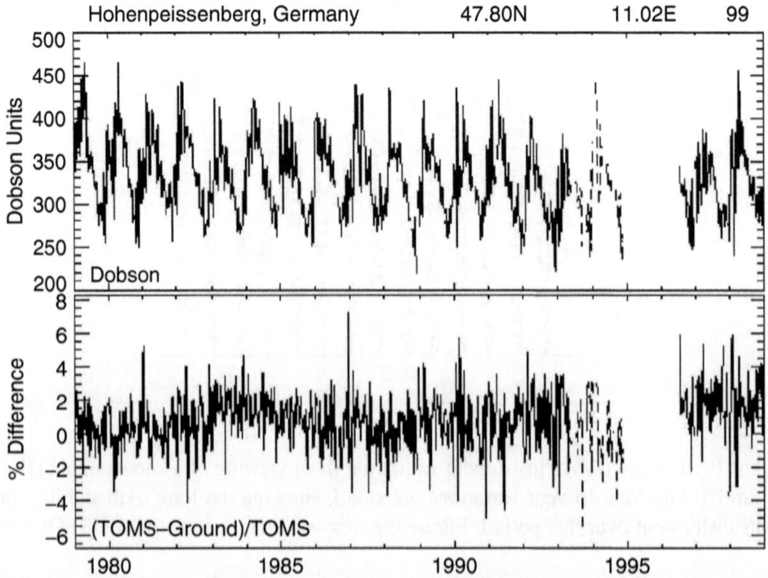

FIGURE 11. Dobson total ozone observations Hohenpeissenberg, Germany, from 1979–2000 in Dobson Units (top panel) and the percentage difference between the collocated TOMS and the Hohenpeissenberg Dobson instrument (bottom panel).

February 1985. Note that the shift is not readily apparent in the upper plot of the raw observations.

Validation of the TOMS observations by ground and the use of TOMS as a transfer standard for the Dobson stations show how validation is an iterative process. Improvements and corrections to the Dobson instruments provide a better comparison to TOMS, which thus provides a better comparison to all of the Dobson instruments. We also see this point in comparisons of the NO_2 instruments that was shown in Figure 6. Mis-registration of the wavelength would be corrected in follow-up comparisons. Validation is an iterative process.

The validation of TOMS provides important general lessons for validation. (1) There must be an onboard calibration that is stable and well-understood. In the case of TOMS, this was accomplished by a good initial design that gave close attention to ground calibration, a long ground calibration effort prior to launch, and the inclusion of calibration lamps and a triple diffuser plate as part of the instrument. (2) There must be in-flight algorithmic calibration and validation techniques. As we've seen with TOMS, the calibration of the radiances from instrument and validation of these radiances is key to its performance. (3) The algorithm conversion from radiances to a total column ozone amount requires accurate physical parameters and models. Such parameters include ozone absorption cross-sections (wavelength and temperature dependence), solar flux measurements, Rayleigh scattering calculations, and the testing of the algorithm using

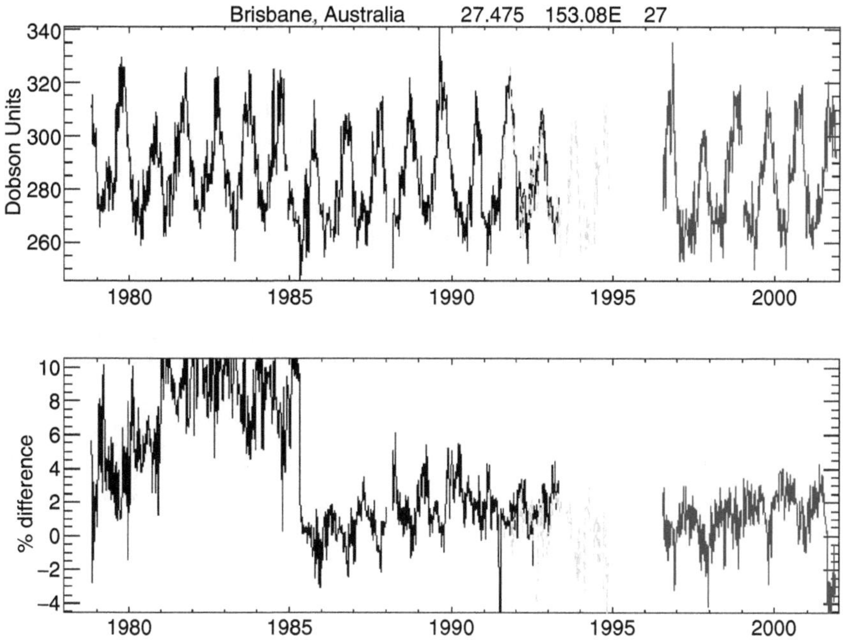

FIGURE 12. Similar to Figure 11. Dobson total ozone observations Brisbane, Australia, from 1979–2000 in Dobson Units (top panel) and the percentage difference between the collocated TOMS and the Brisbane Dobson instrument (bottom panel).

forward models. (4) External validation of TOMS is performed by comparisons to a large amount of data that varies as a function of location, season, and long-term. These external data include Dobson, Brewer, lidar, balloons, microwave, and aircraft in-situ data. The TOMS inter-comparison also includes comparisons to other satellite instruments such as the NOAA SBUV/2, TOMS itself (e.g., between the Nimbus-7 and the Meteor-3 TOMS instruments), Aura / OMI, GOME, and Sciamachy. Large amounts of data increase statistical confidence of comparisons. (5) The comparison of observations from TOMS to the external data is done as an ongoing process that is iterative, and involves detailed exchanges of information amongst scientists.

Noncoincident External Validation

Many validation efforts fail because of the difficulty of co-locating data in both space and time. This is immediately evident when comparing two satellite instruments that measure at different times of the day or with significant distances between observations made at the same time. Most stratospheric validation studies limit coincidence to less than a few hours temporally and less than a few hundred kilometers spatially. Tropospheric validation is even more stringent because the scales of phenomena are smaller (e.g., cloud systems). However, such limits tend to be rather arbitrarily set. In general, validation of coincident observations should be determined by autocorrelation statistics.

Two techniques have been developed to perform validation for non-coincident stratospheric observations. The first technique was developed using potential vorticity and potential temperature comparisons [Lait et al., 1990]. This technique uses potential vorticity (PV) and potential temperature (θ) derived from gridded meteorological data. The PV and θ are interpolated to the observational points. Since PV and θ are conserved on short time scales (30 days) in the lower stratosphere, air parcels observed by different instruments that have identical PV and θ values are "co-located" in a PV-θ coordinate system, and so we can compare observations at widely separated distances. The second technique is trajectory mapping [Morris et al., 1995]. This technique uses both backward and forward trajectories that are performed from the location of a measurement. When these trajectories pass in close proximity to the location of a second observation, we can compare those observations.

Summary

As mentioned at the start, Cal/Val is fundamental to the scientific process. Repeatable and defendable observations provide the foundation of science. The Cal/Val process begins with the design of an instrument. Adequate calibration needs to be engineered into an instrument at its first conception. All elements of an instrument require calibration. These elements include the direct measurement of radiances, the collection of ancillary data, and calibration of algorithms. Internal in-flight cal is a prime requirement for a good instrument. These calibrations must be made against known and relatively unambiguous standards (typically internationally established standards). An instrument should have extensive characterization and pre-flight (and post-flight if possible) calibration.

The Cal/Val process is both continuous and iterative. Initial validation of instruments must be thorough, but long-term validation is equally necessary because of instrument degradation. The importance of climate and ozone trends into the next century points up the critical necessity for continuous Cal/Val over the life of an instrument. In addition, Cal/Val needs to be iterative process between instruments.

Documentation and operating standards are critical for both calibration and validation. This has been rarely mentioned herein, but complete documentation is necessary if the instrument structure and details are to be known. Further, without complete documentation, the measurements can't be repeated or simulated under similar conditions. The Cal/Val efforts must be published in order to permanently establish data set quality for future scientists.

Validation requires both internal and external observations. Internal validation provides an initial test of a data set. External validation requires multiple data sets with lots of match-ups taken over a variety of conditions (i.e., testing the extreme conditions). The external validation (if done between ground or airborne instruments) should be done under blind or double blind conditions. External validation can also be done between non-coincident data using new techniques.

Validation requires communication, collaboration, and publications between scientists. This process of communicating, collaborating, and publishing leads to new ideas for validation, improvements of small errors, and publishing of peer-reviewed articles in the literature for the benefit of the entire scientific community.

References

Diak, G. R., D. S. Kim, M. S. Whippl, X. H. Wu, Preparing for the AMSU, *B. Amer. Met. Soc.*, **73**, 1971–1984, 1992.

Gaffen, D. J., M. A. Sargent, R. E. Habermann, J. R. Lanzante, Sensitivity of tropospheric and stratospheric temperature trends to radiosonde data quality, *J. of Climate*, **13**, 1776–1796, 2000.

Graham, R. J., S. R. Anderson, M. J. Bader, The relative utility of current observation systems to global-scale NWP forecasts, *Q. J. of Roy. Met. Soc.*, **126**, 2435–2460, 2000.

Hoffman, D. J., et al., Intercomparison of UV/visible spectrometers for measurements of stratospheric NO_2 for the Network for the Detection of Stratospheric Changes, *J. Geoph. Res.*, **100**, 16765–16791, 1995.

Hurst, D. F., et al., Construction of a unified, high-resolution nitrous oxide data set for ER-2 flights during SOLVE, *J. Geophy. Res.*, **107**, 8271, doi:10.1029/2001JD000417, 2002.

Ivanov, I., A. Kats, S. Kurnosenko, N. Nash, N. Zaitseva, Instruments and Observing Methods, WMO International Radiosonde Comparison–Phase III, *Report #40, WMO/TD-451*, 1991.

Lait, L. R., et al., Reconstruction of O3 and N2O fields from ER-2, DC-8, and balloon observations. *Geophys. Res. Lett.*, **17**, 521–524, 1990.

Lait, L. R., et al., Non-coincident inter-instrument comparisons of ozone measurements using quasi-conservative coordinates, *Atmos. Chem. Phys.*, **4**, 2345–2352, 2004.

Morris, G. M., et al., Trajectory mapping and applications to data from the Upper Atmosphere Research Satellite, *J. Geophy. Res.*, **100**, 16491–16505, 1995.

Newman, P. A., Antarctic Total Ozone in 1958, *Science*, **264**, 543–546, 1994.

Rigaud, P., B. Leroy, Presumptive evidence for a low value of the total ozone content above Antarctica in September, 1958, *Ann. Geophys.*, **11**, 791–794, 1991.

Tans, P. P., I. Y. Fung, T. Takahashi, Observational constraints on the global atmospheric CO2 budget, *Science*, **247**, 1431–1438, 1990.

Part IV
Output of the Observational Web

Chapter 15

The Sensor Web: A Future Technique for Increasing Science Return

Mark R. Schoeberl and Stephen Talabac

Most current observational systems are set up in a simple way. For example, with satellite sensors, data are collected from a sensor and transmitted to the ground to be processed and distributed to the users. The system is completely passive – the sensor is "on" almost all the time and data processed or assimilated by ground computers. Surface-based networks of radiosondes or air quality monitors operate in a similar way. The sensors (or sensor operators) gather data and relay it to a collection point where it is processed.

The passive data collection process makes sense if there are few sensors and limited goals. But as the number of Earth satellites and surface sensors has increased along with the capability of the sensors, scientists and engineers are rethinking this strategy. For example, scientists are now combining data from different satellite and ground-based sensors, merging these data sets into advanced computer models, and using the results to understand more complex and interrelated phenomenon. For example, to understand aerosols, satellite sensors in the visible, IR, and UV along with ground-based sun photometry and lidar data are needed. Bringing this data together is a difficult data management processes. Furthermore, the amount of data being generated keeps increasing as satellite sensor spectral and spatial resolution increases and as the number of surface-based sites keeps increasing. Scientists and engineers are now seriously considering local data processing so that only the essential information is sent to the ground.

The Sensor Web is an idea about transforming future observing systems that has the potential to significantly increase useful science data return by focusing more sensor power on specific phenomena. If we think of the sensors as distributed nodes with some local data processing and decision-making capability, we can now view the sensor field as a system or web of data gathering elements which can be more generically managed or given broad data gathering goals or perhaps even being directed by prediction models. The sensor web would intelligently and dynamically reconfigure the measurement and information processing states.

To develop the sensor web several minor infrastructure and design changes need to be made. We will need programmable or "smart" taskable sensors, improved spacecraft capability and on-board data processing techniques. We will need better satellite-to-satellite communication links. Why will we need these capabilities? To take advantage of high spatial and spectral resolution of future instrumentation, we will need to be able to collect data over wide swath widths, we need to send this large amount of data to the surface in a single orbital pass, and we will need to transmit or process this data fast enough to make near real-time decisions on observational strategy.

For discussion purposes, we shall talk about three classes of sensor webs. The lowest type system (Type 0) is the completely stove-piped system that describes our current system. Each sensor is independent; the data are simply transmitted to the ground for analysis, the sensor system operation is nearly fixed. At the next level (Type 1) the sensor can react to a changing environment. For example, the sensor can point to a target, or send back a special part of the radiometric spectrum with relatively quick response (hours or minutes). This rapid reaction will allow specialized observations to be made. Experiments with the EO-1 spacecraft have demonstrated the feasibility of this system for viewing fires or volcanic eruptions. For the Type 3 system, we can imagine sensor systems linked together with model calculations (for prediction and analysis) where the sensor system is autonomous or nearly autonomous. For example, we might imagine a predictive model forecasting the development of a Pacific typhoon. The system would automatically take additional observations in the vicinity of the nascent typhoon without intervention. Specialized assets would be deployed including additional balloon launches from near by ships or islands; unmanned aircraft would be flown into the system.

How would the sensor web work? What kind of new autonomy and communication technologies would need to be developed? To explore the potential benefits of sensor web observing strategies for the Earth science community, a number of studies have begun. NASA's EOS "A-Train" constellation (named after members Eos Aqua and Aura, Aqua (http://eos-pm.gsfc.nasa.gov/), Aura (http://eos-aura.gsfc.nasa.gov/) has been used to estimate the requirements (Figure 1). These studies focus on the Aura Troposphere Emission Spectrometer (TES) (Beer et al., 2001) which "follows" 15 minutes behind Aqua with its Moderate Resolution Imaging Spectroradiometer (MODIS) instrument. MODIS is a nadir pointing multispectral imager, it is always "ON", and its 36 visible and infrared bands can be used to detect clouds.

TES is a high-resolution infrared-imaging Fourier transform spectrometer designed primarily to study ozone in the lower atmosphere. Providing spectral coverage in the mid- to thermal IR bands. It is undesirable for TES to make measurements within cloud obscured fields of view. TES is a pointable instrument: it can be commanded to image any target within 45 ° of the local vertical. Test studies on the sensor web use the MODIS data to generate a cloud mask product in real time. Using a predefined list of desired targets, in conjunction with the cloud mask information, the software will generate simulated commands and "point" TES to only those targets that are cloud-free. This demonstrates the value of intelligent data collection measurement techniques. To further refine where within the cloud-free regions TES should point, we are also assessing the potential use of

FIGURE 1. Five spacecraft will comprise the EOS afternoon, or "A Train," constellation shown above. Aqua, launched on May 4, 2002, leads this formation flying spacecraft train. Following Aqua, in the same orbital plane, there are Calipso, Cloudsat, Parasol, and Aura, respectively. All spacecraft are in a sun synchronous, 98.2 degree inclination, 705 Km altitude orbit. Aqua's 1:30PM ascending node will precede Aura's planned 1:38PM ascending node. Orbital characteristics are such that Aqua will lead Aura's position in orbit by approximately 15 minutes.

atmospheric chemistry models whose outputs may provide an additional set of criteria for high science value target selection. Similar "event-driven" and "model-driven" adaptive observation strategies and dynamic measurement techniques may be applied to future sensor web Earth- and space-science missions.

As mentioned above, we can broadly define the Type 3 sensor web as a distributed, organized system of "nodes," interconnected by a communications fabric, that behave as a coherent instrument. The web consists of sensor, computing, and storage nodes. Multiple, heterogeneous, remote sensing and in-situ sensor nodes may be located in space, deployed within the atmosphere (e.g., radiosondes, UAVs), and on or below the Earth's surface (e.g., buoys, autonomous marine craft). Computing nodes, such as data assimilation and weather forecast models and storage nodes, such as intelligent archives [1] complement the sensor nodes. Information from any of these nodes can be used to initiate changes to sensor spatial, spectral, and/or temporal measurement modes. Similarly, a computing node may be reconfigured to change the model's initial conditions, grid scale, or other information processing properties.

The exchange of sensor measurement data (raw or processed) and predictive forecast model information would cause the Type 3 sensor web to adapt and react by initiating new node measurement and information processing states. The potential benefits of this new closed-loop approach include: maximizing the return of only the most useful scientific measurement data; minimizing system response time when monitoring rapidly evolving or transient phenomena; reducing numerical forecast model error growth by performing targeted observations of model-sensitive regions; and improving the system's ability to identify key precursor signatures of environmental phenomena.

The sensor web will probably develop naturally as sensor systems become interconnected and become "smarter." But to include spaceborne systems in the web, planning must begin now, because these systems need a lead time of 10 years or more for technology development and implementation.

References

Beer, R., T. Glavich, D.M. Rider, Tropospheric emission spectrometer for the earth observing system's Aura satellite, *Applied Optics*, **40**(15), May 20, 2001.

Ramapriyan, H, M.G. McConaughy, C. Lynnes, R. Harberts, L. Roelofs, S. Kempler, K. McDonald, Intelligent archive concepts for the future, Proceedings of the ISPRS/Future Intelligent Earth Observing Systems Conference, Novembe,r 2002.

Chapter 16

Fundamentals of Modeling, Data Assimilation, and High-Performance Computing

RICHARD B. ROOD

Introduction

This lecture will introduce the concepts of modeling, data assimilation, and high-performance computing as it relates to the study of atmospheric composition. The lecture will work from basic definitions and will strive to provide a framework for thinking about development and application of models and data assimilation systems. It will not provide technical or algorithmic information, leaving that to textbooks, technical reports, and ultimately scientific journals. References to a number of textbooks and papers will be provided as a gateway to the literature.

The text will be divided into four major sections.

- Modeling
- Data assimilation
- Observing system
- High-performance computing

Modeling

Dictionary definitions of model include

- A work or construction used in testing or perfecting a final product.
- A schematic description of a system, theory, or phenomenon that accounts for its known or inferred properties and may be used for further studies of its characteristics.
- In atmospheric modeling the scientist is generally faced with a set of observations of parameters, for instance, wind, temperature, water, ozone, etc., as well as either the knowledge or expectation of correlated behavior between the different parameters. A number of types of models could be developed to describe the observations. These include:
- Conceptual or heuristic models that outline in the simplest terms the processes that describe the interrelation between different observed phenomena. These

models are often intuitively or theoretically based. An example would be the tropical pipe model of Plumb and Ko [1992], which describes the transport of long-lived tracers in the stratosphere.
- Statistical models that describe the behavior of the observations based on the observations themselves. That is, the observations are described in terms of the mean, the variance, and the correlations of an existing set of observations. Johnson et al. [2000] discuss the use of statistical models in the prediction of tropical sea surface temperatures.
- Physical models that describe the behavior of the observations based on first principle tenets of physics (chemistry, biology, etc.). In general, these principles are expressed as mathematical equations, and these equations are solved using discrete numerical methods. Good introductions to modeling include [Trenberth, 1992; Jacobson, 1998; Randall, 2000].

In the study of geophysical phenomena, there are numerous subtypes of models. These include comprehensive models that attempt to model all of the relevant couplings or interactions in a system and mechanistic models that have one or more parameters prescribed, for instance by observations, and then the system evolves relative to the prescribed parameters. All of these models have their place in scientific investigation, and it is often the interplay between the different types and subtypes of models that leads to scientific advance.

Models are used in two major roles. The first role is diagnostic, in which the model is used to determine and to test the processes that are thought to describe the observations. In this case, it is determined whether or not the processes are well known and adequately described. In general, since models are an investigative tool, such studies are aimed at determining the nature of unknown or inadequately described processes. The second role is prognostic; that is, the model is used to make a prediction. All types of models can be used in these roles.

In all cases the model represents a management of complexity; that is, the scientist is faced with a complex set of observations and their interactions and is trying to manage those observations in order to develop a quantitative representation. In the case of physical models, which are implicitly at the focus of this lecture, a comprehensive model would represent the cumulative knowledge of the physics (chemistry, biology, etc.) that describe the observations. It is tacit that an accurate, comprehensive physical model is the most robust way to forecast; that is, to predict the future.

While models are a scientist's approach to manage and to probe complex systems, today's comprehensive models are themselves complex. In fact, the complexity and scope of models is great enough that teams of scientists are required to contribute to modeling activities. Two consequences of complexity of models are realized in computation and interpretation.

Comprehensive models of the Earth system remain outside the realm of the most capable computational systems. Therefore, the problem is reduced to either looking at component models of the Earth system, i.e., atmosphere, ocean, land, cryosphere, lithosphere, or at models where significant compromises are taken in the representation of processes in order to make them computationally feasible. More challenging, and in fact the most challenging aspect of modeling, is the interpretation of model results. It is much easier to build models than it is to do quantitative evaluation of models and observations.

In order to provide an overarching background for thinking about models it is useful to consider the elements of the modeling, or simulation, framework described in Figure 1. In this framework are six major ingredients. The first are the boundary and initial conditions. For an atmospheric model, boundary conditions are topography and sea surface temperature; boundary conditions are generally prescribed from external sources of information. It is the level of prescription of boundary conditions and the conventions of the particular discipline that determine whether or not a model is termed mechanistic.

The next three items in the figure are intimately related. They are the representative equations, the discrete or parameterized equations, and the constraints drawn from theory. The representative equations are the analytic form of forces or factors that are considered important in the representation of the evolution of a set of parameters. In general, all of the analytic expressions used in atmospheric modeling are approximations; therefore, even the equations the modeler is trying to solve have *a priori* errors. Generally in the construction of a model, only terms that are expected to be important are included in the analytic expressions; that is, the equations are scaled from some more complete representation [see Holton, 2004]. The solution is, therefore, a balance amongst competing forces and tendencies. Most commonly, the analytic equations are a set of nonlinear partial differential equations.

The discrete or parameterized equations arise because it is generally not possible to solve the analytic equations in closed form. The strategy used by scientists is to develop a discrete representation of the equations, which are then solved using numerical techniques. These solutions are, at best, discrete estimates to solutions of the analytic equations. The discretization and parameterization of the analytic equations introduce a large source of error. This introduces another level of balancing in the model; namely, these errors are generally managed through a subjective balancing process that keeps the numerical solution from running away to obviously incorrect estimates.

Boundary Conditions	Emissions, SST, …	ε
Representative Equations	$DA/Dt = P - LA - v/HA + q/H$	ε
Discrete/Parameterize	$(A_{n+\Delta t} - A_n)/\Delta t = \ldots$	$(\varepsilon_d, \varepsilon_n)$
Theory/Constraints	$\partial u_g/\partial z = -(\partial T/\partial y)R/(Hf_0)$	Scale Analysis
Primary Products (*i.e.* A)	$T, u, v, \Phi, H_2O, O_3 \ldots$	$(\varepsilon_b, \varepsilon_v)$
Derived Products (F(A))	Pot. Vorticity, v^*, w^*, …	Consistent

FIGURE 1. Simulation framework.

While all of the terms in the analytic equation are potentially important, there are conditions or times when there is a dominant balance between, for instance, two terms. An example of this is thermal wind balance in the middle latitudes of the stratosphere [see Holton, 2004]. It is these balances, generally at the extremes of spatial and temporal scales, which provide the constraints drawn from theory. Such constraints are generally involved in the development of conceptual or heuristic models. If the modeler implements discrete methods that consistently represent the relationship between the analytic equations and the constraints drawn from theory, then the modeler maintains a substantive scientific basis for the interpretation of model results.

The last two items in Figure 1 represent the products that are drawn from the model. These are divided into two types: primary products and derived products. The primary products are variables such as wind, temperature, water, ozone – parameters that are most often explicitly modeled; that is, an equation is written for them. The derived products are often functional relationships between the primary products; for instance, potential vorticity. A common derived product is the balance, or the budget, of the different terms of the discretized equations. The budget is then studied, explicitly, on how the balance is maintained and how this compares with budgets derived directly from observations. In general, the primary products can be directly evaluated with observations and errors of bias and variability estimated. If attention has been paid in the discretization of the analytic equations to honor the theoretical constraints, then the derived products will behave consistently with the primary products and theory. They will have errors of bias and variability, but they will behave in a way that supports scientific scrutiny.

In order to explore the elements of the modeling framework described above, the following continuity equation for a constituent, A, will be posed as the representative equation. The continuity equation represents the conservation of mass for a constituent and is an archetypical equation of geophysical models. Brasseur and Solomon [1986] and Dessler [2000] provide good backgrounds for understanding atmospheric chemistry and transport. The continuity equation for A is:

$$\partial A/\partial t = -\nabla \cdot \mathbf{U}A + M + P - LA - n/HA + q/H \tag{1}$$

where

— A is some constituent
— **U** is velocity → “resolved” transport, “advection”
— M is “Mixing” → “unresolved” transport, parameterization
— P is production
— L is loss
— n is “deposition velocity”
— q is emission
— H is representative length scale for n and q
— t is time
— ∇ is the gradient operator

Attention will be focused on the discretization of the resolved advective transport. Figures 2 and 3 illustrate the basic concepts. On the left of the figure a mesh has been laid down to cover the spatial domain of interest. In this case it is a rectangular mesh. The mesh does not have to be rectangular, uniform, or orthogonal. In fact, the mesh can be unstructured or can be built to adapt to the features that are being modeled. The choice of the mesh is determined by the modeler and depends upon the diagnostic and prognostic applications of the model [see Randall, 2000]. The choice of mesh can also be determined by the computational advantages that might be realized.

Points can be prescribed to determine location with the mesh. In Figure 2, both the advective velocity and the constituent are prescribed at the center of the cell. In Figure 3, the velocities are prescribed at the center of the cell edges, and the constituent is prescribed in the center of the cell. There are no hard and fast rules about where the parameters are prescribed, but small differences in their prescription can have huge impact on the quality of the estimated solution to the equation, i.e., the simulation. The prescription directly impacts the ability of the model to represent conservation properties and to provide the link between the analytic equations and the theoretical constraints [see Rood, 1987; Lin 2004]. In addition, the prescription is strongly related to the stability of the numerical method; that is, the ability to represent any credible estimate at all.

A traditional and intuitive approach to discretization is to use differences calculated across the expanse of the grid cell to estimate partial derivatives. This is the foundation of the finite-difference method, and finite differences appear in one form or another in various components of most models. Differences can be calculated from a stencil that covers a single cell or weighted values from neighboring cells can be used. From a numerical point of view, the larger the stencil, the more cells that are used, the more accurate the approximation of the derivative. Spectral methods, which use orthogonal expansion functions to estimate the derivatives, essentially use information from the entire domain. While the use of

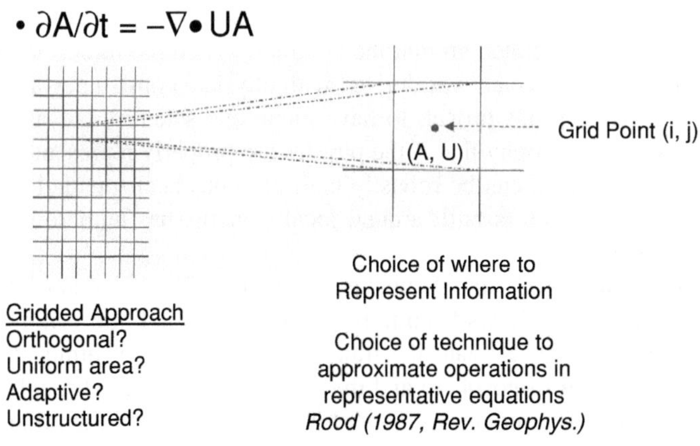

FIGURE 2. Discretization of resolved transport.

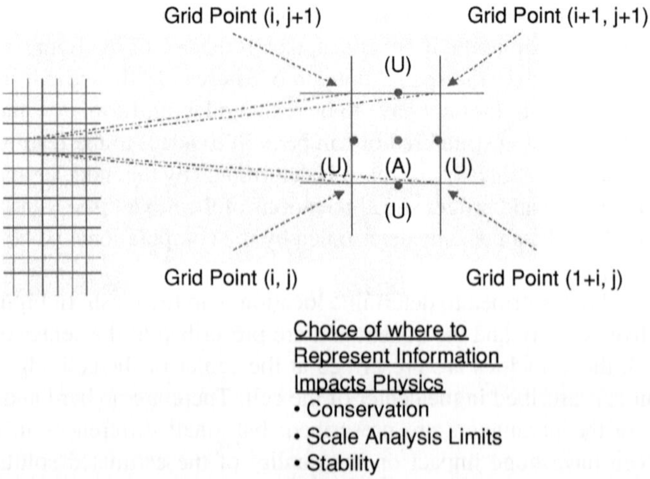

FIGURE 3. Discretization of resolved transport.

a large stencil generally increases the accuracy of the estimate of the partial derivatives, it also increases the computational cost and means that discretization errors are correlated across large portions of the domain.

The use of numerical techniques to represent the partial differential equations that represent the model physics is a straightforward way to develop a model. However, there are many approaches to discretization of the dynamical equations that govern geophysical processes [Jacobson, 1998; Randall, 2000]. Given that these equations are, in essence, shared by many scientific disciplines, there are sophisticated and sometimes similar developments in many different fields. One approach that has been recently adopted by several modeling centers is described in Lin [2004]. In this approach the cells are treated as finite volumes and piecewise continuous functions are fit locally to the cells. These piecewise continuous functions are then integrated around the volume to yield the forces acting on the volume. This method, which was derived with physical consistency as a requirement for the scheme, has proven to have numerous scientific advantages. The scheme uses the philosophy that if the physics are properly represented, then the accuracy of the scheme can be robustly built on a physical foundation. In addition, the scheme, which is built around local stencils, has numerous computational advantages.

Douglass et al. [2003] and Schoeberl et al. [2003] have demonstrated the improved quality that follows from implementation of the finite volume scheme. In their studies they investigate the transport and mixing of atmospheric constituents in the upper troposphere and the lower stratosphere. Through a combination of analysis of observations, a hierarchy of models, and the relationship to theoretical constraints, they demonstrate that both the horizontal and vertical transport is better represented with the finite volume scheme. Further, their comparisons of experiments using winds from several data assimilation systems to calculate transport establish the importance of physical consistency in the representation of budgets of the constituent continuity equation.

Data Assimilation

The definition of assimilation from the dictionary is

- To incorporate or absorb; for instance, into the mind or the prevailing culture

For Earth science, assimilation is the incorporation of observational information into a physical model. Or more specifically:

- Data assimilation is the objective melding of observed information with model-predicted information.

Returning to the discussion of model types in the previous section, assimilation rigorously combines statistical modeling with physical modeling; thus, formally connecting the two approaches. Daley [1991] is the standard text on data assimilation. Cohn [1997] explores the theory of data assimilation and its foundation in estimation theory. Swinbank et al. [2003] is a collection of tutorial lectures on data assimilation. Assimilation is difficult to do well, easy to do poorly, and its role in Earth science is expanding and sometimes controversial.

Figure 4 shows elements of an assimilation framework that parallels the modeling elements in Figure 1. The concept of boundary conditions remains the same; that is, some specified information at the spatial and temporal domain edges. Of particular note, the motivation for doing data assimilation is often to provide the initial conditions for predictive forecasts.

Data assimilation adds an additional forcing to the representative equations of the physical model; namely, information from the observations. This forcing is formally added through a correction to the model that is calculated, for example, by [see Stajner et al., 2001]:

$$(\boldsymbol{O}P_f\boldsymbol{O}^T + R)x = A_o - \boldsymbol{O}A_f \tag{2}$$

The terms in the equation are as follows:

— A_o are observations of the constituent
— A_f are model forecast, simulated, estimates of the constituent

Model		Data
Emissions, SST, ...	ε	Boundary Conditions
$DA/Dt = P - LA - v/HA + q/H$	ε	$(\boldsymbol{O}P^f\boldsymbol{O}^T + R)x = A_o - \boldsymbol{O}A_f$
$(A_{n+\Delta t} - A_n)/\Delta t = \ldots$	ε	Discrete/Error Modeling
$\partial u_g/\partial z = -(\partial T/\partial y)R/(Hf_0)$	Scale Analysis	Constraints on Increments
$A_i \equiv T, u, v, \Phi, H_2O, O_3 \ldots$	$(\varepsilon_h, \varepsilon_v)$	$(\varepsilon_h, \varepsilon_v)$ reduced
Pot. Vorticity, v*, w*, ...	Consistent	Inconsistent

FIGURE 4. Assimilation framework.

— $\boldsymbol{O}$ is the observation operator
— P_f is the error covariance function of the forecast
— R is the error covariance function of the observations
— x is the innovation that represents the observation-based correction to the model
— ^{T} is the matrix transform operation

The observation operator, $\boldsymbol{O}$, is a function that maps the parameter to be assimilated onto the spatial and temporal structure of the observations. In its simplest form, the observation operator is an interpolation routine. Often, however, it is best to perform assimilation in observation space, and in the case of satellite observations the measurements are radiances. Therefore, the observation operator might include a forward radiative transfer calculation from the model's geophysical parameters to radiance space. While this is formally robust, in practice, it is sometimes less than successful because of loss of intuitiveness and computational problems. Therefore, initial experiments with assimilation are often most productively undertaken using retrieved geophysical parameters.

The error covariance functions, P_f and R, represent the errors, respectively, of the information from the forecast model and the information from the observations. This explicitly shows that data assimilation is the error-weighted combination of information from two primary sources. These error covariance functions are generally not well-known. From first principles, the error covariance functions are prohibitive to calculate. Hence, models are generally developed to represent the error covariance functions. Stajner et al. [2001] show a method for estimating the error covariances in an ozone assimilation system.

Parallel to the elements in the simulation framework (Figure 1), discrete numerical methods are needed to estimate the errors as well as to solve the matrix equations in Eq. (2). How and if physical constraints from theory are addressed is a matter of both importance and difficulty. Often, for example, it is assumed that the increments of different parameters that are used to correct the model are in some sort of physical balance. For instance, wind and temperature increments might be expected to be in geostrophic balance. However, in general, the data insertion process acts like an additional physics term in the equation and contributes a significant portion of the budget. This, explicitly, alters the physical balance defined by the representative equations of the model. Therefore, there is no reason to expect that the correct geophysical balances are represented in an assimilated data product. This is contrary to the prevailing notion that the model and observations are "consistent" with one another after assimilation.

The final two elements in Figure 4 are, again, the products. In a good assimilation the primary products, in general the prognostic variables, are well estimated. That is, both the bias errors and the variance errors are reduced. However, the derived products are likely to be physically inconsistent because of the nature of the corrective forcing added by the observations. These errors are often found to be larger than those in self-determining model simulations. Molod et al. [1996] and Kistler et al. [2001] provide discussions on the characteristics of the errors associated with primary and derived products in data assimilation systems. The nature of the errors described in these papers is consistent with errors in present-day assimilation systems.

A schematic of an assimilation system is given in Figure 5. This is a sequential assimilation system where a forecast is provided to a statistical analysis algorithm that calculates the merger of model and observational information. In this example, errors are specified based on external considerations and methods. There is a formal interface between the statistical analysis algorithm and the model prediction which performs a quality assessment of the information prior to the merger; this quality control algorithm will be discussed more fully below. The figure shows, explicitly, two input streams for the observations. The first of these streams represent the observations that will be assimilated with the model prediction. The other input stream represents observations that will not be assimilated. This second stream of observations could be, for example, a new type of observation whose error characteristics are being determined relative to the existing assimilation system. The second stream might also represent an ancillary data set that is being used in quality control decisions. This type of monitoring function finds many applications, and data assimilation systems are excellent tools for determining anomalies in input data streams.

In Figure 4 the products of the assimilation were classified as primary and derived estimates of geophysical parameters. The following classification of products describes the collective information from the data assimilation system. These are indicated in Figure 5 and listed below:

- Analysis: The analysis is the merged combination of model information and observational information. The analysis is the best estimate of the state of the system based on the optimization criteria and error estimates.
- Forecast/simulation: The forecast/simulation is a model run that starts from an initial condition defined by the analysis. For some amount of time this model

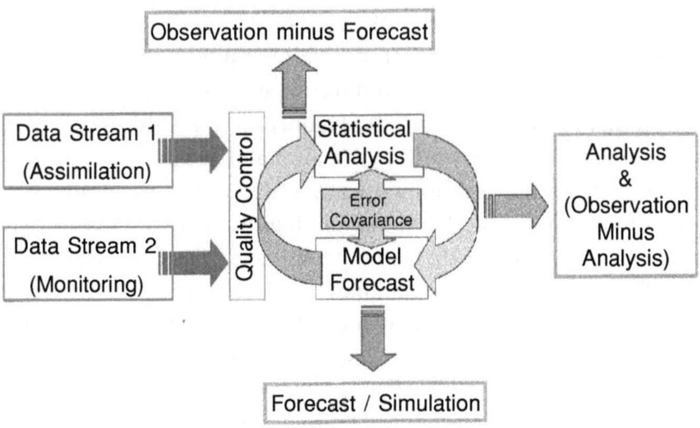

FIGURE 5. Schematic of data assimilation system.

run is expected to represent the state of the system with some deterministic accuracy. For this case the model run is a forecast. After a certain amount of time the model run is no longer expected to represent the particular state of the system; though, it might represent the average state. In this case the model run is simply a simulation that has been initialized with a realistic state estimate at some particular time.

- Observation minus forecast increment: The observation minus forecast increment, often the O-F, gives a raw estimate of the agreement of the forecast information with the observation information prior to assimilation. Usually, a small O-F increment indicates a high-quality forecast, and O-F increments are used as a primary measure of the quality of the assimilation. O-F increments are exquisitely sensitive to changes in the system and are the primary quantity used for monitoring the stability and quality of the input data streams. Study of the O-F is useful for determining the spatial and temporal characteristics of some model errors.
- Observation minus analysis increment: The observation minus analysis increment represents the actual changes to the model forecast that are derived from the statistical analysis algorithm. Therefore, they represent in some bulk sense the error weighted impact of the O-F increments. If the assimilation system weighs the observations heavily relative to the forecast, then the observation minus analysis increments will have significant differences relative to the O-F increments. The opposite is also true; if the model information is weighed more heavily than the observational information then there will be little change to the O-F increments. If either of these extremes are realized the basic assumptions of the assimilation problem need to be reconsidered.

As suggested earlier, the specification of forecast and model error covariances and their evolution with time is a difficult problem. In order to get a handle on these problems it is generally assumed that the observational errors and model errors are unbiased over some suitable period of time, e.g., the length of the forecast between times of data insertion. It is also assumed that the errors are in a Gaussian distribution. The majority of assimilation theory is developed based on these assumptions, which are, in fact, not valid assumptions. In particular, when the observations are biased, there would the expectation that the actual balance of geophysical terms is different from the balance determined by the assimilation. Furthermore, since the biases will have spatial and temporal variability, the balances determined by the assimilation are quite complex. Aside from biases between the observations and the model prediction, there are biases between different observation systems of the same parameters. These biases are potentially correctible if there is a known standard of accuracy defined by a particular observing system. However, the problem of bias is a difficult one to address and perhaps the greatest challenge facing assimilation [see Dee and da Silva, 1998]

As a final general consideration, there are many time scales represented by the representative equations of the model. Some of these time scales represent balances that are achieved almost instantly between different variables. Other time scales are long, important to, for instance, the general circulation, which will determine the distribution of long-lived trace constituents. It is possible in assimilation to produce a very accurate representation of the observed state variables and those variables that are balanced on fast time scales. On the other hand, improved estimates in the state variables are found, at least sometimes, to be associated with degraded estimates of those features determined by long time scales. Conceptually, this can be thought of as the impact of bias propagating through the physical model. With the assumption that the observations are fundamentally accurate, this indicates errors in the specification of the physics that demand further research.

Data assimilation has had dramatic impacts in the improvement of weather forecasts. In other applications the benefits of assimilation have been more difficult to realize. Therefore, scientists need to determine the appropriateness of assimilation or using assimilated data products in their studies. The list below provides goals of the assimilation of ozone data. These goals are examples that can be extended to the assimilation of other geophysical parameters.

- Mapping: There are spatial and temporal gaps in the ozone observing system. A basic goal of ozone assimilation is to provide vertically resolved global maps of ozone.
- Short-term ozone forecasting: There is interest in providing operational ozone forecasts in order to predict the fluctuations of ultraviolet radiation at the surface of the earth [Long et al., 1996].
- Chemical constraints: Ozone is important in many chemical cycles. Assimilation of ozone into a chemistry model provides constraints on other observed constituents and helps to provide estimates of unobserved constituents.
- Unified ozone data sets: There are several sources of ozone data with significant differences in spatial and temporal characteristics as well as their expected error characteristics. Data assimilation provides a potential strategy for combining these data sets into a unified data set.
- Tropospheric ozone: Most of the ozone is in the stratosphere, and tropospheric ozone is sensitive to surface emission of pollutants. Therefore, the challenges of obtaining accurate tropospheric ozone measurements from space are significant. The combination of observations with the meteorological information provided by the model offers one of the better approaches available to obtain global estimates of tropospheric ozone.
- Improvement of wind analysis: The photochemical time scale for ozone is long compared with transport timescales in the lower stratosphere and upper troposphere. Therefore, ozone measurements contain information about the wind field that might be obtained in multivariate assimilation.
- Radiative transfer: Ozone is important in both longwave and shortwave radiative transfer. Therefore, accurate representation of ozone is important in the radiative transfer calculations needed to extract (retrieve) information from many satellite instruments. In addition, accurate representation of ozone has

the potential to impact the quality of the temperature analysis in multivariate assimilation.

- Observing system monitoring: Ozone assimilation offers an effective way to characterize instrument performance relative to other sources of ozone observations as well as the stability of measurements over the lifetime of an instrument.
- Retrieval of ozone: Ozone assimilation offers the possibility of providing more accurate initial guesses for ozone retrieval algorithms than are currently available.
- Assimilation research: Ozone (constituent) assimilation can be productively approached as a univariate linear problem. Therefore, it is a good framework for investigating assimilation science; for example, the impact of flow-dependent covariance functions.
- Model and observation validation: Ozone assimilation provides several approaches to contribute to the validation of models and observations.

Some of the goals mentioned above can be meaningfully addressed with the current state of the art. Others cannot. It is straightforward to produce global maps of total column ozone, which can be used in, for instance, radiative transfer calculations. The use of ozone measurements to provide constraints on other reactive species is an application that has been explored since the 1980s [see Jackman et al., 1987] and modern data assimilation techniques potentially advance this field. The impact of ozone assimilation on the meteorological analysis of temperature and wind, and hence improvement of the weather forecast, is also possible. The most straightforward impact would be on the temperature analysis in the stratosphere. The improvement of the wind analysis is a more difficult challenge and confounded by the fact that where improvements in the wind analyses are most needed, the tropics, the ozone gradients are relatively weak. The use of ozone assimilation to monitor instrument performance and to characterize new observing systems is currently possible and productive [see Stajner et al., 2004]. The improvement of retrievals using assimilation techniques to provide ozone first guess fields that are representative of the specific environmental conditions is also an active research topic. The goal of producing unified ozone data sets from several instruments is of little value until bias can be correctly accommodated in data assimilation. This final topic will be discussed more fully below.

There has been much written in the assimilation literature about the various approaches to the assimilation algorithm and the use of assimilated data sets in many types of applications. There has been relatively little written about the quality control of the observations; that is, the interface between the observations and the model predicted data sets. The successes or failures of data assimilation systems can however be directly related to decisions made in the quality control [see Dee et al., 2001]. A simple description of quality control is given in Figure 6.

On the left side of the figure are three sources of observational information, two satellites and one nonsatellite source. These three sources of observations might represent a nadir viewing temperature sounder, a limb viewing temperature sounder, and radiosonde temperatures. Even if perfectly accurate, each of these observing systems would provide different measurements because of the

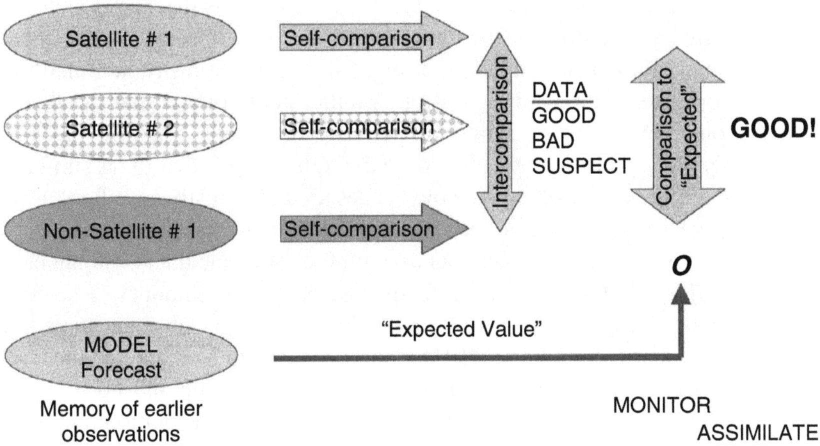

FIGURE 6. Quality control: Interface to the observations.

sampling characteristics of the instrument. Quality control of the observations might proceed as follows. Each type of observation is likely to come with some information about the anticipated quality of the data. This information might indicate whether or not an observation is far outside the expected value based on the previous performance of the instrument, or alternatively, that the instrument was in an observing mode known to have low reliability. Further investigation of observations that are flagged, as say, suspicious, might reveal that there is a region of geographical consistency; that is, a region of suspicious data. This region could represent a meaningful geophysical feature, perhaps a developing storm, or it might represent a regional contamination of the observations, perhaps clouds or an erupting volcano.

If the investigation of the observation suggests that the observations might be of geophysical interest, then intercomparison with other types of observations can confirm this suggestion. Since the different types of observations might have different environmental sensitivities, the identification of a regional anomaly in all of the observation types would add weight to favor the inclusion of the suspicious data in the assimilation. Finally, the model prediction can be brought into the decision-making process. The model is an estimate of the projected value of the observation, and the observation minus forecast information is a sensitive indicator of information. If the model suggests there is a developing storm, then inclusion of the data is likely to better represent the forecast of that storm. If the model does not show a storm, but all three types of observations suggest that there is a storm, then the analysis will reflect the storm, and an otherwise missed feature will be correctly forecast.

Quality control decisions are difficult and can have significant impact on the assimilation quality [Dee at al., 2001]. It is intuitive that a handful of observations taken near clouds that represent a real developing storm will have much more impact than additional observations in clear skies where persistence is expected. In a scientific investigation of the data system, the data rejected by the quality control demand further investigation. They could reflect an instrument malfunction or an operator or transmission error. Another possibility is that the field of view is contaminated by a cloud, or perhaps a volcanic eruption has been detected. Finally, systematic rejection of data might suggest that the assimilation system is drifting because the error covariances are not robustly specified, or that a new geophysical phenomenon, perhaps a trend, is being measured.

Figure 7 shows an example from an assimilation of ozone data [Wargan et al., 2005]. In this example, there are two satellite instruments, the Solar Backscattered Ultraviolet/2 (SBUV) and the Michelson Interferometer for Passive Atmospheric Sounding (MIPAS). SBUV is a nadir sounder and measures very thick layers with the vertical information in the middle and upper stratosphere. MIPAS is a limb sounder with much finer vertical resolution and measurements extending into the lower stratosphere. SBUV also measures total column ozone, which is assimilated in all experiments. The results from three assimilation experiments are shown through comparison with an ozonesonde profile. Ozonesondes were not assimilated into the systems; therefore, these data provide an independent measure of performance.

There are several attributes to be noted in Figure 7. The quality of the MIPAS-only (+ SBUV total column) assimilation is the best of those presented. This suggests that the vertical resolution of MIPAS instrument is having a large impact. Even though the MIPAS observations are assimilated only above 70 hPa, the assimilation captures the essence of the structure of the ozone profile down to 300 hPa. This indicates that the model information in the lower stratosphere and upper troposphere is geophysically meaningful. Further, the model is effectively distributing information in the horizontal between the satellite profiles. The comparison with the SBUV-only (+ SBUV total column) assimilation shows

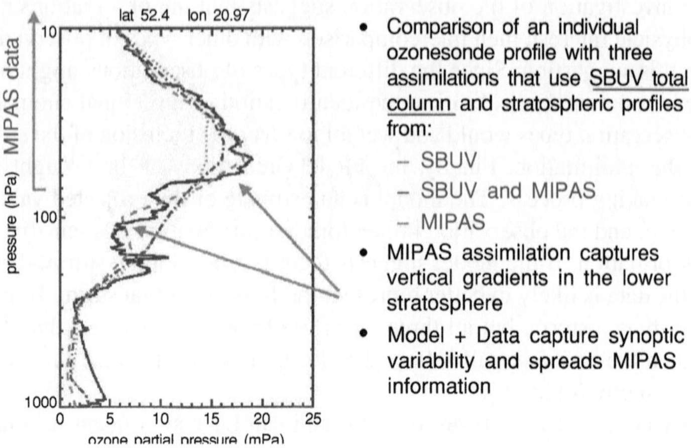

FIGURE 7. MIPAS ozone assimilation.

that the thick-layered information of the SBUV observations, even in combination with the model information, does not represent the ozone peak very well. This impacts the quality of the lower stratospheric analysis as the column is adjusted to represent the constraints of the total ozone observations. Finally, from first principles, the combined SBUV and MIPAS assimilation might be expected to be the best; this should have the maximum amount of information. This is not found to be the case and suggests that the weights of the various error covariances and the use of the observations can be improved. The optimal balance of nadir and limb observations is not straightforward, and these experiments reveal the challenges that need to be addressed when multiple types of instruments are used in data assimilation.

The Observing System

The example of ozone assimilation discussed in Figure 7 hints at the attention that needs to be paid to the characteristics of the instruments in the observing system. There are questions of accuracy and bias between different observations of the same parameter. Similarly, there are questions of how different instruments measure the variability of geophysical parameters. When different, but correlated, parameters are measured, for example, ozone and temperature, there is the question of how measurement errors fit into the correlated behavior of the parameters. Data assimilation brings yet another level of attention to the observations; namely, the specific characteristics of the observing system: how do the footprints of two instruments impact their use? What is the depth of the averaging kernel? How are limb scanning and nadir measurements used together? How are sparse, high-quality localized observations used? Should dense observations be sampled or averaged prior to use? How is information from integrated quantities, such as outgoing longwave radiation, used? One approach to answering these questions is to transform the model variables into the same space as the observations through the observation operator see Eq. (2). As noted, above, the rigor of this strategy is often challenged by the reality of the implementation and the interpretation of the problem so that the approach is not straightforward or successful.

The observational data used in data assimilation are often broken into three categories, listed below:

- Conventional Data: Conventional data are those data that are, in principle, from the presatellite era. More generally, conventional data are nonsatellite data. These data include surface observations and balloon soundings, as well as ship and aircraft observations. Many of these data are collected operationally, and they are a critical part of meteorological (and oceanographic and land-surface) assimilation systems. There are also data collected in research missions that are of exceptional quality that find their use in data assimilation, often as independent validation data. Though nonsatellite data might be classified as conventional, the optimal use of these observations requires careful attention to the details of the data systems. Lait [2002] provides an interesting examination of the radiosonde network and the impact that instruments from different countries and manufacturers have on the quality of the analysis.

- Operational Satellite Data: Operational satellite data are those data taken routinely to support national weather prediction centers. These data are taken and processed in real-time and distributed around the world. Because of these real-time requirements, there are limitations on the resolution of the observations, the size of the data sets, and the sophistication of retrieval algorithms and forward models. Historically, the calibration and stability of operational satellites has been of secondary importance; they are essentially calibrated by the conventional data as communicated through the assimilation system.
- Research Satellite Data: Research satellite data are those instruments that have been launched for scientific exploration or technology proof of concept. Research satellites measure parameters that have not been measured before or measure parts of the environment that have been adequately sampled. For example, measurements of carbon dioxide from space would be a new, important measurement; measurements of water and temperature in the planetary boundary layer would resolve these parameters in a region of especial importance. Research satellites might also seek to resolve traditional satellite measurements, temperature or ozone, with higher accuracy, higher spectral resolution, or higher horizontal and vertical resolution than operational satellites.

The entire suite of observations must be considered in a comprehensive data assimilation activity. Even if a research group is considering the impact of a particular instrument on a particular problem, usually a foundational data assimilation system is required that adequately considers the primary variables from the conventional and operational satellite data systems.

Though small in number, in meteorological data assimilation the conventional data are very important to the quality of the analysis and the forecast. This is perhaps due to the high quality of these observations as well as a good knowledge of the error characteristics. Alternatively, the assimilation systems might be implicitly tuned to these observations because of their long heritage and the legacy of adding new data types onto the pre-existing data system. Scientific investigation and re-investigation of the existing conventional and operational data systems would be an interesting and potentially productive research activity.

By shear quantity, the operational satellite data far outnumber the conventional data. The satellite data assures high-quality global analyses, and satellite data have been essential to the continual improvement of weather forecasts. Research satellite data not only improve the quality of analyses relative to the pre-existing data system, but extend the analysis to new parameters and new domains; for instance, constituents and chemistry, land-surface, ocean prediction, etc. Because of the great cost of research instruments, there is increasing use of research observations at operational centers to assure that the research instruments benefit society.

The use of research satellites in operation applications greatly increases the complexity of the assimilation process. For many years conventional data and operational satellites worked under tight conventions that assured both common data formats and a small number of data centers that supplied all of the nations of the world. However, with the use of research observations, centers and scientists must interact with many specialized data processing organizations and use a multitude of data formats. In addition, both the operational and research satellites are contributing to a vast increase in the number of available observations. This increase in the number of observations, projected to be as much as six orders of magnitude between the years 2000 and 2010, will overwhelm data systems and computational capabilities unless new techniques are developed for the selection and use of data to be assimilated. Data usage provides, yet another, major challenge and is the focus of much of the research and development at operational centers.

An accurate operational data assimilation system provides the ideal interface between scientists and observing systems. There are many possibilities for developing an adaptive observing system – i.e., a tunable sensor web. In fact, such a notion provides one of the strategies for addressing the computational challenges of the projected enormous increase in data volume. The data assimilation system could target which of the observations from the satellites might be expected to have the greatest impact on a particular application, e.g., the forecast. Then only those data might be selected for evaluation or retrieval and possible inclusion in the data assimilation. In the future, it might be possible to direct the satellite to particular parts of the Earth and to target and take, only, those observations expected to have the greatest impact.

There are two natural places for the data assimilation system to provide interfaces to an adaptive observing system (see Figure 5). The first is the forecast, where particular features might be identified several hours or days in advance, then targeted. The second is with the observation minus forecast (O-F) increments. Large increments indicate places where the expected values from the forecast agree poorly with the observations. There are many possible reasons for disagreement, and one possibility is a region of high uncertainty, perhaps due to a poorly simulated developing system. Extra retrievals or targeted observations from any data platform could verify or refute the existence of such a developing system.

Finally, there are two basic strategies of observing. One is targeted observing of features or processes of special interest. The other is sampling or surveying of the entire domain. Both of these strategies are essential ingredients of scientific investigation. It is not a matter of one or the other. Robust assimilation depends on the existence of an observing system that adequately samples the domain. With this foundation, the idea of targeted observations to investigate those features that are not adequately represented by routine sampling makes sense.

High-Performance Computing

Two aspects of computing will be discussed. First, the characteristics of the computational problem that distinguish Earth-science modeling and assimilation will be discussed. Second, the attributes that influence the increasing need for computational resources will be discussed.
The dictionary definition of a computer is

- A device that determines a result by mathematical or logical operations.
  High-performance computing describes a niche of computing that is associated with those platforms able to address the most demanding calculations. Since all aspects of computational technology (processor speed, memory, storage, network speed, etc.) show an exponential increase in capability as a function of time, the technical specifications of high-performance computing is also a function of time. High-performance computing is also called supercomputing and high-end computing [NAS, 2005].
  There are a number of potential definitions or descriptions for high performance computing. These include
- Computing that is 1 – 2 orders of magnitude beyond that available from the state-of-the-art desktop environment, or alternatively, beyond that which can be acquired by a well-funded principal investigator.
- "The class of fastest and most powerful computers available," from Landau and Fink [1993]

Two important attributes are common to applications requiring high-performance computers. The first is that multiple computational processors must be gathered together and made to operate on a single image of the application software in order to achieve acceptable time to solution. Current practices in high-performance computing centers would suggest that applications that require approximately 64 processors on a single job would be termed high-performance. The second attribute is that special attention must be paid to the management of memory during the run time of the application.

These attributes highlight that high-performance computing is not simply an issue of hardware, but one also of software. In order to make effective use of a high-performance computer the scientist must have high-performance software. High-performance software must be able to scale to multiple processors; that is, the software must be able to utilize additional processors efficiently. As additional processors are added, the efficiency of each added processor is reduced because of communications overhead. There is a point at which adding more processors does little to increase the performance of the application. At the heart of efficient scaling is the management of memory. If the information needed by the processors can be kept at ready access to the processors, then efficient scaling can be maintained. This suggests that memory bandwidth; that is, how fast does information transfer from memory to the processor is an important aspect of a high-performance computer. In many applications that involve fluid flow, the physics of the problem require that information from one processor be communicated to other processors. This provides a formidable challenge to the scientific programmer, which is specifically related to the memory architecture of the hardware. This brings the need to add specialized computer programmers to the teams that aspire to comprehensive modeling and data assimilation activities. NAS [2001] provides an excellent examination of the problems writing scaleable software for climate models and the interaction of hardware and software.

There are two competing approaches by computer manufacturers to build high-performance computers. The first is to build specialized platforms that are anchored around custom processors, custom memory architecture, and custom communication interconnects amongst the processors. This adds significant cost to the computational platform. Since high-performance computing is a small part of the market, these tightly integrated platforms do not provide cost-performance numbers that appeal to the majority of the market. The second strategy, therefore, is to build high-performance computers out of components that are commercially available. This takes advantage of the exponential growth of increasing component capability. However, this requires that the computer companies build the environment that connects these components together – components that have not generally been developed to work together. This, again, adds significant cost to the computational platform. Further, this second approach pushes more of the work to the scientific programmer developing high-performance software. The issues of high-performance computing, its role in science, and their link to market factors are discussed in NAS [2005].

The need for high-performance computing is driven by both the requirement that scientific investigation requires a certain level of computational completeness to be productive and the requirement that the time to solution allows the products of the computation to be useful in their application. The workload suggested by these requirements falls into two natural categories – capability computing and capacity computing. These are described in Figure 8. Capability is defined by the maximum number of processors that can be focused, efficiently,

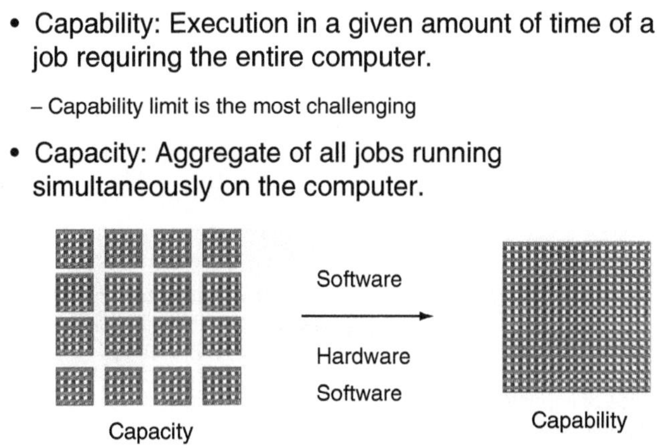

FIGURE 8. Computational capacity & capability.

on a single application. Capability is generally driven by a demand for increasing realism and comprehensiveness in a calculation, or a requirement that a product be produced in a given time segment (*i.e.*, real-time requirements). Capacity generally describes the execution of many applications that individually do not require highest capability. An example of capability computing would be a high-resolution, deterministic weather forecast; an example of capacity computing would be an ensemble of low-resolution forecasts to develop probabilistic information. Both capability and capacity computing are important to Earth-science modeling and assimilation. Sometimes, however, scientists are limited in capability experiments because of the expense and difficulty of writing high-performance software.

A heuristic example that demonstrates the communication issues of high-performance computing can be made from consideration of a group of people who need to make a series of transactions. If each person can make their transaction without negotiation with other people, for instance, buying their own lunch, then a group of people is well served by having a number of cashiers. However, if the group ordered their lunches together and need to negotiate with each other over the amount that each individual needs to pay, and further, requires the cashier to participate in the execution of their negotiation, then having more than one cashier is of little benefit. The computational problem is, therefore, defined not only by the number of transactions (calculations), but also by the amount of negotiation (communications) required. Figure 9 uses the format of Figures 2 and 3 to illustrate this point for the modeling of a hurricane. Assuming that the grid points are now associated with a certain subset of the processors (Figure 8), then information from one processor is needed from other processors to determine the internal dynamics of the hurricane. In addition, the path that the hurricane follows connects information from a series of grid points and processors. It is intuitive that the choice of grid, the specification of variables, and the selection of a discretization routine will impact the computational performance. While this example demonstrates the need for communications in a particular problem, other applications, for example, land-surface assimilation, might only have weak

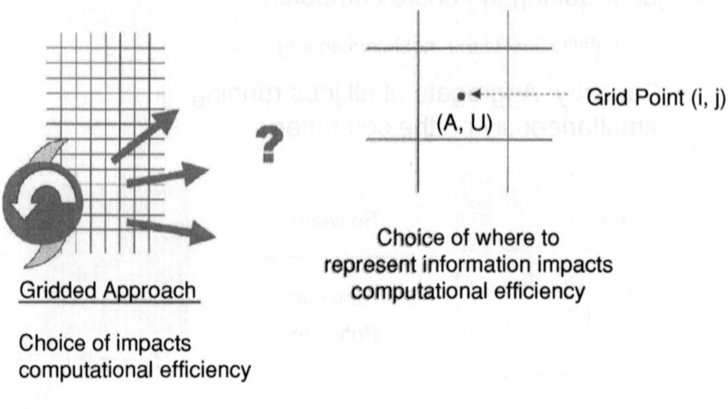

FIGURE 9. Requirement for parallel communication.

requirements for communications. Therefore, loosely connected computation platforms might be adequate.

Finally, there are several aspects of the modeling and assimilation problem that stress computational systems and push capability requirements. The common ones in modeling are increased resolution, improved physics, inclusion of new processes, and integration and concurrent execution of Earth-system components that are normally run separately – that is, coupled models. Often, real-time needs define capability requirements. When considering data assimilation the computational requirements become much more challenging. Often the computational characteristics of the statistical analysis are defined as a function of the number and distribution of the observations; therefore, the increasing number of observations could be computationally crippling. Advanced assimilation techniques often involve iterative cycling between the model and the statistical analysis routine, increasing the computational burden. The increasing diversity of data sources and the use of research observations in assimilation place tremendous demands on networks and data systems. The computational details of modeling, statistical analysis, and quality control are quite different. As with the construction of a state-of-the art model or data assimilation system, balanced cost considerations need to be made in the computational aspects of the problem – both software and hardware. This requires the scientist and the science manager to constantly consider the tension between the reduction of the problem to its component parts and the unification of those parts into a system.

Summary

This lecture introduced the fundamental ideas that a scientist needs to understand when building or using models in Earth-science research. Rather than focusing on technical aspects of modeling and data assimilation, the lecture focused on a number of underlying principles. These principles, if adhered to, will allow the models and model products to be used in quantitative, data-driven research. This goes beyond simple comparison of models and observations and using their similarity as a measure of worth.

With regards to standalone models in the absence of data assimilation, it was emphasized that the underlying physics should be well represented. This requires special attention to the physics as using accurate numerical techniques does not guarantee physical consistency. Data assimilation was introduced as adding a forcing term to the model that is a correction based on observations. This additional forcing term changes the balance of forces. Therefore, budgets calculated from assimilated data are not expected to be robust for geophysical applications. For both modeling and data assimilation, it is much more difficult to quantitatively analyze and interpret the results than it is to develop new modules and components. Few scientists do this analysis well, and students are challenged to learn existing techniques and to develop new techniques.

With regard to data assimilation, the importance of the observing system was emphasized. This requires monitoring of the observing system and vigorous attention to quality control. It also requires attention to the details of the instrumentation, for example, the observational technique. Scientific investigation of the observing system was encouraged. The importance of bias in data assimilation was also discussed. The presence of bias lies at the foundation of the physical consistency of assimilated data sets. While data assimilation has had a number of outstanding successes, these issues of bias and physical consistency require scientists to consider the appropriateness of data assimilation to their particular problem.

The attention to the observing system brought out the changing nature of the observing system. Specifically, the observing system is becoming more diverse and data volumes are increasing rapidly. This requires the efforts of many scientists and new computational techniques to utilize these new observations effectively. Data assimilation systems provide a natural link between scientists and the observing system, including the possibility of adaptive observing systems.

Finally, the computational aspects of modeling and assimilation were discussed. Comprehensive activities that address the entire Earth system remain beyond the most capable computers. Special challenges come from the fact that many computations are required (transaction) and communication is required between the computations (negotiation). In addition, the computational issues of faced when embracing the data systems are often different than those usually considered in standalone modeling activities. Computational considerations must be incorporated in the development of data assimilation systems, and again, they need to also address issues of physical consistency.

The end-to-end data assimilation system must have balance. There is little benefit developing components to a high state of accuracy or performance if there are other weaknesses in the system. Experience suggests that future progress will be most effectively realized through the use of new data types and improving the representation of parameterized physics in models. With these efforts, bias might be addressed in a fundamental way. If the bias problem is not addressed, then there are intrinsic limitations to the problems appropriately addressed by data assimilation.

References

Brasseur, G., S. Solomon, *Aeronomy of the Middle Atmosphere,* D. Reidel Publishing Company, 452 pp, 1986.

Cohn, S. E., An introduction to estimation theory, *J. Met. Soc. Japan,* **75** (1B), 257–288, 1997.

Daley, R., *Atmospheric Data Analysis,* Cambridge University Press, 457 pp, 1991.

Dee, D. P., A. da Silva, Data assimilation in the presence of forecast bias, *Q. J. Roy. Meteor. Soc.*, **124** (545), 269–295, Part A, 1998.

Dee, D. P., L. Rukhovets, R. Todling, A. M. da Silva, J. W. Larson, An adaptive buddy check for observational quality control, *Q. J. Roy. Meteor. Soc.*, **127** (577), 2451–2471, Part A, 2001.

Dessler, A.E., *The Chemistry and Physics of Stratospheric Ozone*, Academic Press, 214 pp, 2000.

Douglass, A.R., M.R. Schoeberl, R.B. Rood, S. Pawson, Evaluation of transport in the lower tropical stratosphere in a global chemistry and transport model, *J. Geophys. Res.*, **108**, Art. No. 4259, 2003.

Holton, J.R., *An Introduction to Dynamic Meteorology,* Elsevier Academic Press, 535 pp, 2004.

Jacobson, M.Z., *Fundamentals of Atmospheric Modeling,* Cambridge University Press, 672 pp, 1998 -(2nd Ed, to appear, 2005).

Jackman C.H., P.D. Guthrie, J.A. Kaye, An intercomparison of nitrogen-containing species in Nimbus 7 LIMS and SAMS data, *J. Geophys. Res.*, **92**, 995–1008, 1987.

Johnson, S.D., D.S. Battisti, E.S. Sarachik, Empirically derived Markov models and prediction of tropical Pacific sea surface temperature anomalies, *J. Climate,* **13**, 3–17, 2000.

Kistler R., E. Kalnay, W. Collins, et al., The NCEP-NCAR 50-year Reanalysis: Monthly means CD-ROM and documentation, *Bull. Amer. Meteor. Soc.* **82**, 247–267, 2001.

Lait, L.R., Systematic differences between radiosonde measurements, *Geophys. Res. Lett.*, **29**, Art. No. 1382, 2002.

Landau, R.H., and P.J. Fink, *A Scientist's and Engineer's Guide to Workstations and Supercomputers,* John Wiley and Sons, 416 pp, 1993.

Lin, S.J., A "vertically Lagrangian" finite-volume dynamical core for global models, *Mon. Wea. Rev.*, **132**, 2293–2307, 2004.

Long, C.S., A.J. Miller, H.T. Lee, J.D. Wild, R.C. Przywarty, D. Hufford, Ultraviolet index forecasts issued by the National Weather Service, *Bull. Amer. Meteorol. Soc.*, **77**, 729–748, 1996.

Molod, A., H.M. Helfand, L.L. Takacs, The climatology of parameterized physical processes in the GEOS-1 GCM and their impact on the GEOS-1 data assimilation system, *J. Climate,* **9**, 764–785, 1996.

NAS, *Improving the Effectiveness of U.S. Climate Modeling*, National Academy Press, Washington, DC, 142 pp, 2001 (http://books.nap.edu/catalog/10087.html).

NAS, *Getting Up to Speed: The Future of Supercomputing,* National Academy Press, S.L. Graham, M. Snir, and C.A. Paterson (eds.), Washington, DC, 308 pp, 2005 (http://books.nap.edu/catalog/11148.html).

Plumb, R.A., M.K. W. Ko, Interrelationships between mixing ratios of long lived stratospheric constituents, *J. Geophys. Res.*, **97**, 10145–10156, 1992.

Randall, D.A. (Ed.), *General Circulation Model Development: Past, Present, and Future,* Academic Press, 807 pp, 2000.

Rood, R.B., Numerical advection algorithms and their role in atmospheric transport and chemistry models, *Rev. Geophys.*, **25**, 71–100, 1987.

Schoeberl, M.R., A.R. Douglass, Z. Zhu, S. Pawson, A comparison of the lower stratospheric age-spectra derived from a general circulation model and two data assimilation systems, *J. Geophys. Res.*, **108**, Art. No. 4113, 2003.

Stajner, I., L.P. Riishojgaard, R.B. Rood, The GEOS ozone data assimilation system: Specification of error statistics, *Q. J. R. Meteorol. Soc.*, **127**, 1069–1094, 2001.

Stajner, I., N. Winslow, R.B. Rood, S. Pawson, Monitoring of observation errors in the assimilation of satellite ozone data, *J. Geophys. Res.*, **109**, Art. No. D06309, 2004.

Swinbank, R., R. Shutyaev, W.A. Lahoz (eds.), *Data Assimilation for the Earth System,* NATO Science Series: IV: Earth and Environmental Sciences, 26, Kluwer, 388 pp, 2003.

Trenberth, K.E. (ed.), *Climate System Modeling,* Cambridge University Press, 788 pp, 1992.

Wargan, K., I. Stajner, S. Pawson, R.B. Rood, and W. –W. Tan, Assimilation of ozone data from the Michelson interferometer for passive atmospheric sounding, *Quart. J. Royal. Met. Soc.*, to appear, 2005.

Chapter 17

Inverse Modeling Techniques

DANIEL J. JACOB

We often seek to determine the state of a system by making measurements of the observable manifestations of that system, and using our physical understanding of the system to relate these observations to the state. In a complex system, such as the atmosphere, the number of variables defining the state is very large. Useful characterization requires a model driven by a limited number of state variables. Let us define an n-dimensional subset of these state variables that we wish to quantify through observations. We assemble this subset into a *state vector* **x**. All other state variables in the model are lumped into a *parameter vector* **b**. We now make m measurements of the observable variables that we assemble in an *observation vector* **y**. Our model provides a *forward* relationship $\mathbf{y} = \mathbf{F}(\mathbf{x}, \mathbf{b})$ between **x** and **y**. By inverting this model, analytically or numerically, we can obtain **x** given **y**. In general we have errors on the measurements, errors in the physical model, errors in the model parameters, as well as prior knowledge of the state vector, that all need to be taken into account in the inversion. An *inverse model* describes the structure by which measurements of the observation vector **y** are used to obtain optimal constraints on the state vector **x**.

Consider, for example, the problem of quantifying surface fluxes of CO_2 on a (latitude × longitude × month) grid. The fluxes on that grid represent a state vector x. We make observations of atmospheric CO_2 concentrations from a network of sites, representing an observation vector y. We use a CTM as forward model to gives us y = F(x,b). The model parameter vector b includes meteorological variables and any characteristics of the CO_2 flux (such as diurnal variability) that are simulated in the model but not resolved in the state vector. By inverting this CTM we obtain an estimate of x given y. This is called a top-down constraint on the surface fluxes. It has uncertainties due to errors in the observations and in the CTM. In addition, we cannot ignore independent information on what x should be based on our knowledge of the processes determining the fluxes (such as fuel combustion statistics, land type data bases, etc.). We refer to this information as a bottom-up constraint; it represents our *a priori* knowledge before making the observations of y. After making the observations, we use the inverse model combine the top-down and bottom-up constraints and obtain an optimal a posteriori estimate of the surface fluxes. We often call this *a posteriori* estimate the retrieval.

As another example, consider the problem of constructing a complete 3D field of CO_2 concentrations over the globe on the basis of limited measurements of concentrations at isolated points and scattered times. Such a construction may be useful, for example, to assess the consistency of CO_2 measurements made from different platforms. We can define the state vector $\mathbf{x}(t)$ as the 3D ensemble of gridded CO_2 concentrations at time t, and $\mathbf{y}$ as the vector of observations available over the time interval $[t\text{-}\Delta t, t]$. The forward model is a CTM initialized with $\mathbf{x}(t\text{-}\Delta t)$ and providing a *forecast* for time t. This forecast represents our *a priori* information for $\mathbf{x}(t)$, to be corrected on the basis of the observations over $[t\text{-}\Delta t, t]$. The task is similar to the previous example but here our state vector will in general be very large and evolve in time, while the observation vector at any given time is relatively sparse. We refer to this type of inverse model application as *data assimilation.*

Bayes' Theorem

Bayes' theorem provides the general foundation for inverse models. Consider a pair of vectors $\mathbf{x}$ and $\mathbf{y}$. Let $P(\mathbf{x})$, $P(\mathbf{y})$, $P(\mathbf{x}, \mathbf{y})$ represent the corresponding probability distribution functions *(pdfs)*, so that the probability of $\mathbf{x}$ being in the range $[\mathbf{x}, \mathbf{x} + \mathbf{dx}]$ is $P(\mathbf{x})\mathbf{dx}$, the probability of $\mathbf{y}$ being in the range $[\mathbf{y}, \mathbf{y} + \mathbf{dy}]$ is $P(\mathbf{y})\mathbf{dy}$, and the probability of $(\mathbf{x}, \mathbf{y})$ being in the range $[\mathbf{x}, \mathbf{x} + \mathbf{dx}, \mathbf{y}, \mathbf{y} + \mathbf{dy}]$ is $P(\mathbf{x}, \mathbf{y})\mathbf{dxdy}$. Let $P(\mathbf{y}|\mathbf{x})$ represent the pdf for y when x is assigned to a certain value. We have

$$P(\mathrm{y}|\mathrm{x}) = \frac{P(\mathrm{x}, \mathrm{y})}{P(\mathrm{x})} \tag{1}$$

Similarly, let $P(\mathbf{x}|\mathbf{y})$ represent the pdf for $\mathbf{x}$ when $\mathbf{y}$ is assigned to a certain value:

$$P(\mathrm{x}|\mathrm{y}) = \frac{P(\mathrm{x}, \mathrm{y})}{P(\mathrm{y})} \tag{2}$$

Eliminating $P(\mathbf{x}, \mathbf{y})$ from (1) and (2) we obtain Bayes' theorem:

$$P(\mathrm{x}|\mathrm{y}) = \frac{P(\mathrm{y}|\mathrm{x})\,P(\mathrm{x})}{P(\mathrm{y})} \tag{3}$$

The relevance to the inverse problem is readily apparent. If P(x) is the pdf of the state vector x before the measurement is made (that is, the *a priori* pdf), and P(y|x) is the pdf of the observation vector $\mathbf{y}$ given the true value for x, then P(x|y) is the *a posteriori* pdf for the state vector reflecting the information from the measurement. The optimal or maximum *a posteriori* (MAP) solution for x is given by the maximum of P(x|y), that is, the solution to $\nabla_x P(\mathbf{x}|\mathbf{y}) = \mathbf{0}$ where ∇_x is the the gradient operator in the state vector space. The probability function $\mathbf{P}$(y) in the denominator is independent of x, and we can view it merely as a normalizing factor to ensure that $\int_0^\infty P(\mathrm{x}|\mathrm{y})\,\mathbf{dx} = 1$. It plays no role in determining the MAP solution and we generally ignore it.

Inverse Problem for Scalars

Application of Bayes' theorem to obtain the MAP solution is simplest and easiest to understand using scalars. Consider a source releasing a species *X* to the atmosphere with an emission flux *x*. We have an *a priori* estimate of $x_a \pm \sigma_a$ for the value of *x*, where σ_a^2 is the variance of the error on the estimate. For example, if *X* is emitted from a power plant, the *a priori* information would be based on knowledge of the type and amount of fuel being burned in that plant, any emission control equipment, etc. We set up a sampling site to measure the concentration of *X* downwind of the source. We measure a concentration $y \pm \sigma_i$ where σ_i^2 is the variance of the instrument error. We then use a CTM to give us a relationship between *x* and *y* as

$$y = F(x) \pm \sigma_m \tag{4}$$

where σ_m^2 is the variance of the CTM error. Let us assume that the CTM relationship between *x* and *y* is linear in the range of interest. Let us further assume that the instrument and CTM errors are uncorrelated so that the corresponding variances are additive. The measured concentration *y* is then related to the *true* source *x* by

$$y = kx \pm \sigma_\varepsilon \tag{5}$$

where the coefficient *k* is obtained from the CTM, and σ_ε^2 is the variance of the *observational error* defined as the sum of the instrumental and CTM errors:

$$\sigma_\varepsilon^2 = \sigma_i^2 + \sigma_m^2 \tag{6}$$

The observational error includes a contribution from the forward model error, so it is not purely from "observations". Think of it as the error in the *observing system* designed to place constraints on the state vector.

After making the measurement, we seek an improved estimate $\hat{x}$ of *x* that optimally accommodates the top-down constraint from the measurement and the bottom-up constraint from the *a priori*. We use Bayes' theorem. Assuming Gaussian error distributions, we have

$$P(x) = \frac{1}{\sigma_a\sqrt{2\pi}} \exp[-\frac{(x - x_a)^2}{2\sigma_a^2}] \tag{7}$$

$$P(y|x) = \frac{1}{\sigma_\varepsilon\sqrt{2\pi}} \exp[-\frac{(y - kx)^2}{2\sigma_\varepsilon^2}] \tag{8}$$

Applying Bayes' theorem (3) and ignoring the normalizing terms that are independent of *x*, we obtain

$$P(x|y) \exp[-\frac{(x - x_a)^2}{2\sigma_a^2} - \frac{(y - kx)^2}{2\sigma_\varepsilon^2}] \tag{9}$$

Finding the maximum value for $P(x|y)$ is equivalent to finding the minimum in the *cost function* $J(x)$:

$$J(x) = \frac{(x - x_a)^2}{\sigma_a^2} + \frac{(y - kx)^2}{\sigma_\varepsilon^2} \tag{10}$$

which is a least-squares cost function weighted by the variance of the error in the individual terms. It is called a χ^2 cost function, and *J(x)* as formulated in Eq. (10) is called the χ^2 statistic.

The optimal estimate $\hat{x}$ is the solution to $\partial J / \partial x = 0$, which is straightforward to obtain analytically:

$$\hat{x} = x_a + g(y - kx_a) \tag{11}$$

where *g* is a *gain factor* given by

$$g = \frac{k\sigma_a^2}{k^2\sigma_a^2 + \sigma_\varepsilon^2} \tag{12}$$

In (11), the second term on the right-hand side represents the correction to the *a priori* on the basis of the measurement *y*. The gain factor is the sensitivity of the retrieval to the observation: $g = \hat{x}\partial/\partial y$. We see from (12) that the gain factor depends on the relative magnitudes of σ_a and σ_ε/k. If $\sigma_a << \sigma_\varepsilon/k$, then $g \rightarrow 0$ and $\hat{x} \rightarrow x_a$; the measurement is useless because the observational error is too large. If by contrast $\sigma_a >> \sigma_\varepsilon/k$, then $g \rightarrow 1/k$ and $x \rightarrow y/k$; the measurement is so precise that it constrains the solution without recourse to the *a priori* information.

We can also express the retrieval $\hat{x}$ in terms of its proximity to the true solution *x*. Let ε (with expected value σ_ε) describe the error on the measurement; we have

$$y = kx + \varepsilon \tag{13}$$

Replacing in Eq. (11), we obtain

$$\hat{x} = ax + (1 - a)x_a + g\varepsilon \tag{14}$$

where $a = gk$ is an *averaging kernel* describing the relative weights of the *a priori* x_a and the true value *x* in contributing to the retrieval. The averaging kernel represents the sensitivity of the retrieval to the true state: $a = \partial\hat{x} / \partial x$. The gain factor is now applied to the observational error in the third term on the right-hand side. We see from Eq. (12) that the averaging kernel simply weights the errors σ_a and σ_ε/k. In the limit $\sigma_a >> \sigma_\varepsilon/k$, then $a \rightarrow 1$ and the *a priori* does not contribute to the solution. Do we then approach the true solution? Not necessarily, because of the third term $g\,\varepsilon$ in Eq. (14) with expected value $g\,\sigma_\varepsilon$. Since in the above limit $g \rightarrow 1/k$, we obtain in fact $\hat{x} \rightarrow y / k$ as derived previously. The error on the retrieval is then defined by the observational error. We call $(1\text{-}a)x_a$ the *smoothing error* since it limits the ability of the retrieval to obtain solutions departing from the *a priori*, and we call $g\,\sigma_\varepsilon$. the *retrieval error.*

We can derive a general expression for the retrieval error by starting from Eq. (9) and expressing it in terms of a Gaussian distribution for the error in $(x\text{-}\hat{x})$. We thus obtain a form in $\exp[-(x - \hat{x})^2 / 2\hat{\sigma}^2]$ where σ^2 is the variance of the error in the *a posteriori* $\hat{x}$. The calculation is laborious but straightforward, and yields

$$\frac{1}{\sigma^2} = \frac{1}{\hat{\sigma}_a^2} + \frac{1}{(\sigma_\varepsilon / k)^2} \tag{15}$$

Notice that the *a posteriori* error is always less than the *a priori* and observational errors, and tends toward one of the two in the limiting cases that we described.

Let us now consider a situation where our single measurement y is not satisfactory in constraining the solution. We could remediate this problem by making m measurements y_i, each adding a term to the cost function (10). Assuming the same observational error variance for each measurement:

$$J(x) = \frac{(x - x_a)^2}{\varepsilon_a^2} + \sum_m \frac{\langle y - kx \rangle^2}{\sigma_\varepsilon^2 / m} \tag{16}$$

We can re-express *J(x)* as

$$J(x) = \frac{(x - x_a)^2}{\sigma_a^2} + \frac{\langle y - kx \rangle^2}{\sigma_\varepsilon^2 / m} \tag{17}$$

where σ_ε^2 / m is the variance of the error on the mean value $<y - kx>$ (this is the *central limit theorem)*. By increasing *m*, we could thus approach the true solution. This works only if (1) the observational error is truly random, (2) the measurements have uncorrelated observational errors. With regard to (1), it is critical to establish if there is any systematic error (that is, *bias*) in the measurement. In the presence of bias, no number of measurements will allow convergence to the true solution; the bias will be propagated through the gain factor and correspondingly affect the retrieval. Calibration of the measuring instrument and of the CTM is thus essential. With regard to (2), instrumental errors (as from photon-counting) are often uncorrelated; however, the CTM errors rarely are. For example, two successive measurements at a site may sample the same air mass and thus be subject to the same CTM transport error. It is thus important to determine the *error correlation* between the different measurements. This error correlation can best be described by assembling the measurements in a vector and constructing the *observational error covariance matrix* with elements cov(y_i, y_j). Dealing with error correlations, and also dealing with multiple sources, requires that we switch to a vector-matrix notation in our formulation of the inverse problem. We do so in the next section.

We have assumed in the above a linear forward model $y = F(x) = kx$. What if the forward model is not linear? We can still calculate an MAP value for $\hat{x}$ as the minimum in the cost function (10), where we replace kx by the nonlinear form $F(x)$. The error in this MAP solution is not Gaussian though, so Eq. (15) would not apply. An alternative is to linearize the forward model around x_z as $k = \frac{\partial y}{\partial x}(x_a)$ to obtain an initial guess of $\hat{x}$, and then iterate on the solution by recalculating $k = \frac{\partial}{\partial x}(\hat{x})$. As we will see, the latter is the only practical solution as we move from scalar to vector space.

Generalized Linear Inverse Problem

Let us now consider the problem of a state vector **x** of dimension n on which we seek information from observations assembled into an observation vector **y** of dimension m. We have an *a priori* estimate $\mathbf{x_a}$ for the state vector with *error covariance matrix* $\mathbf{S_a}$ *(nxn)*. The error covariance matrix has as diagonal elements the error variances of the individual elements of $\mathbf{x_a}$, and as off-diagonal elements the error covariances of the different elements of $\mathbf{x_a}$. The error variance for element $x_{a,i}$ of vector $\mathbf{x_a}$ is defined as the expected value of $(x_i - x_{a,,i})^2$ when x_i is sampled over its *a priori* pdf $P(x_i)$ (see section 6.2) and the error covariance for the pair $(x_{a,i}, x_{a,j})$ is defined as the expected value of the product $(x_i - x_{a,i})(x_j - x_{a,j})$ where x_i and x_j are sampled over their respective *a priori* pdfs. We express the error covariance matrix in compact mathematical form as $\mathbf{S}_a = E\ \{(\mathbf{x} - \mathbf{x_a})\ (\mathbf{x} - \mathbf{x_a})^T\}$ where E is the expected value operator. For example, if we estimate a uniform 50% error on the individual elements of $\mathbf{x_a}$ with no correlation between the errors on the different elements, then the diagonal elements of $\mathbf{S_a}$ are $0.25x_{a,i}^{\ 2}$ and the off-diagonal elements are all zero.
We use a linear forward model to relate **x** to **y**:

$$\mathbf{y} = \mathbf{K}\mathbf{x} + \varepsilon \tag{18}$$

where K is a mxn Jacobian matrix with terms $k_{ij} = \partial y_i \,/\, \partial x_j$ calculated from the forward model, and ε is an observational error vector with error covariance matrix $\mathbf{S}_\varepsilon = E\ \{\varepsilon\varepsilon^T\}$ of dimension m × m. Included in $\mathbf{S}_\varepsilon$ are all the sources of error that would prevent the linear forward model If the forward model relating **x** to **y** is not linear, then it should be linearized about the *a priori* estimate $\mathbf{x_a}$ using a Taylor expansion. The linearization is then an additional source of error to be included in $\mathbf{S}_\varepsilon$.
Application of Bayes' theorem to this inversion problem follows the same approach as earlier, but the algebra is more complicated. The equivalents of (7), (8), and (9) are

$$-2\ln P(\mathbf{x}) = (\mathbf{x} - \mathbf{x_a})^T \mathbf{S_a}^{-1} (\mathbf{x} - \mathbf{x_a}) + c_1 \tag{19}$$

$$-2\ln P(\mathbf{y}|\mathbf{x}) = (\mathbf{y} - \mathbf{K}\mathbf{x})^T \mathbf{S}_\varepsilon^{-1} (\mathbf{y} - \mathbf{K}\mathbf{x}) + c_2 \tag{20}$$

$$-2\ln P(\mathbf{x}|\mathbf{y}) = (\mathbf{x} - \mathbf{x_a})^T \mathbf{S_a}^{-1} (\mathbf{x} - \mathbf{x_a}) + (\mathbf{y} - \mathbf{K}\mathbf{x})^T \mathbf{S}_\varepsilon^{-1} (\mathbf{y} - \mathbf{K}\mathbf{x}) + C_3 \tag{21}$$

where c_1, c_2, c_3 are constants. Again the maximum *a posteriori* (MAP) solution involves determining the value of **x** that yields the maximum value of $P(\mathbf{x}|\mathbf{y})$, which is equivalent to finding the minimum in the scalar-valued cost function $J(\mathbf{x})$:

$$J(\mathbf{x}) = (\mathbf{x} - \mathbf{x_a})^T \mathbf{S_a}^{-1} (\mathbf{x} - \mathbf{x_a}) + (\mathbf{y} - \mathbf{K}\mathbf{x})^T \mathbf{S}_\varepsilon^{-1} (\mathbf{y} - \mathbf{K}\mathbf{x}) \tag{22}$$

The solution is

$$\hat{\mathbf{x}} = \mathbf{x_a} + \mathbf{G}(\mathbf{y} - \mathbf{K}\mathbf{x_a}) \tag{23}$$

where the *gain matrix* $G = \partial\hat{x}/\partial y$ is given by

$$G = S_a K^T (K S_a K^T + S_\varepsilon)^{-1} \quad (24)$$

or equivalently:

$$G = (K^T S_\varepsilon^{-1} K + S_a^{-1})^{-1} K^T S_\varepsilon^{-1} \quad (25)$$

The error covariance matrix $\hat{\mathbf{S}}$ on the retrieval $\hat{\mathbf{x}}$ can be calculated as before by rearraging the right-hand side of (21) as $(\mathbf{x} - \hat{\mathbf{x}})^T \hat{\mathbf{S}}^{-1} (\mathbf{x} - \hat{\mathbf{x}})$:

$$\hat{S} = (K^T S_\varepsilon^{-1} K + S_a^{-1})^{-1} \quad (26)$$

Note the similarity of (23)–(24) to (11)–(12) and (26) to (15). When the state vector is time-varying so that (23) and (25) are used repeatedly to update the state vector at successive times, the method is often called the Kalman filter.
As earlier, we can express the departure of the retrieval from the true solution in terms of an averaging kernel matrix $\mathbf{A} = \mathbf{GK}$:

$$x = Ax + (In - A)Xa + G\varepsilon \quad (27)$$

where $\mathbf{A}$ represents the sensitivity of the retrieval to the true state: $\mathbf{A} = \partial\hat{x}/\partial x$. In Eq. (26), $(\mathbf{I}_n - \mathbf{A})\mathbf{x}_a$ represents the smoothing error and $\mathbf{G}\boldsymbol{\sigma}_\varepsilon$ represents the retrieval error. We can express $\mathbf{A}$ in terms of the error covariance matrices. By substituting $\mathbf{A} = \mathbf{GK}$ into Eq. (6.25) we obtain

$$A = I - \hat{S} S_a^{-1} \quad (28)$$

which is a convenient way to compute the averaging kernel matrix once $\hat{\mathbf{S}}$ has been obtained.

Retrieval Capabilities of an Observing System

The ability of an observing system to constrain the different components of the chosen state vector can be diagnosed by inspecting the successive rows of the averaging kernel matrix. Row i of A has as elements $\partial\hat{x}_i / \partial x_j$ for $j = 1,\ldots n$. Ideally one would like $\partial\hat{x}_i / \partial x_i = 1$ and $\partial\hat{x}_i / \partial x_j = 0$ for $j \neq i$. If $\partial\hat{x}_i / \partial x_j \neq 0$ *that means that changes in the true value of* x_j *will project into the retrieved value for* x_i.

One is also interested in determining the number of pieces of information on a state vector provided by the observing system, commonly referred to as the number of degrees of freedom for signal (DOFS). This is discussed in detail by Rodgers. The DOFS is the trace of the averaging kernel matrix. Another good way to characterize the DOFS is with the renormalized Jacobian matrix $\hat{K} = S_\varepsilon^{-\frac{1}{2}} K S_a^{\frac{1}{2}}$. The DOFS is given by the number of eigenvalues of $\hat{K}^T\hat{K}$ that are greater than unity, and the corresponding eigenvectors of $\hat{K}^T\hat{K}$ represent an orthonormal basis in state space that is constrained by the observing system. Although it might seem that this would be a good way to choose optimally the state vector, the eigenvectors of $\hat{K}^T\hat{K}$ represent in general complicated patterns; the first few can often be recognized to describe important features resolved by the observing system, but the following ones are typically difficult to interpret.

Index

MIX
Papier aus verantwortungsvollen Quellen
Paper from responsible sources
FSC® C105338

If you have any concerns about our products,
you can contact us on
ProductSafety@springernature.com

In case Publisher is established outside the EU,
the EU authorized representative is:
Springer Nature Customer Service Center GmbH
Europaplatz 3, 69115 Heidelberg, Germany

Printed by Libri Plureos GmbH
in Hamburg, Germany